Theatre Pedagogy in the Era of Climate Crisis

This volume explores whether theatre pedagogy can and should be transformed in response to the global climate crisis.

Conrad Alexandrowicz and David Fancy present an innovative re-imagining of the ways in which the art of theatre, and the pedagogical apparatus that feeds and supports it, might contribute to global efforts in climate protest and action. Comprised of contributions from a broad range of scholars and practitioners, the volume explores whether an adherence to aesthetic values can be preserved when art is instrumentalized as protest and considers theatre as a tool to be employed by the School Strike for Climate movement. Considering perspectives from areas including performance, directing, production, design, theory and history, this book will prompt vital discussions which could transform curricular design and implementation in the light of the climate crisis.

Theatre Pedagogy in the Era of Climate Crisis will be of great interest to students, scholars and practitioners of climate change and theatre and performance studies.

Conrad Alexandrowicz, MFA, is an associate professor in the Department of Theatre at the University of Victoria, where he teaches movement for actors. He had a lengthy career as a performer and creator of physical theatre, and he has been a dancer, choreographer, playwright, director and producer. His first book, *Acting Queer: Gender Dissidence and the Subversion of Realism*, was published in February 2020.

David Fancy, PhD, is full professor and chair in the Department of Dramatic Arts in the Marilyn I. Walker School of Fine and Performing Arts at Brock University. He publishes on philosophy, technology, environmentalism, disability and performance and has a creative practice as a playwright and director of theatre, opera and circus. He is editor of a website on the subject of diversities and actor training.

"This collection is an important contribution to the urgent conversations concerning the climate crisis and how theatre artists and scholars can address these pressing topics through activism, Indigenous knowledge systems, performance pedagogies, and performance."

– Rachel Bowditch, Professor, Director of Graduate Studies for Theatre and Dance, Arizona State University, USA

"I am overwhelmed by a deep and personal gratitude for this book. The collection succeeds in employing performance to address the failure of imagination that is the root cause of the climate crisis. It offers a healing pedagogy to dress the wounds of individuality, and the consumer/dominator culture, towards a human/nature, mind/body unity, dissolving into a great wide belonging. Deeply philosophical and refreshingly practical, this is an essential book for enacting an equitable, survivable and thrive-able future for all life."

– Beth Osnes, Associate Professor of Theatre and Performance Studies, University of Colorado, USA

Research and Teaching in Environmental Studies

This series brings together international educators and researchers working from a variety of perspectives to explore and present best practice for research and teaching in environmental studies.

Given the urgency of environmental problems, our approach to the research and teaching of environmental studies is crucial. Reflecting on examples of success and failure within the field, this collection showcases authors from a diverse range of environmental disciplines including climate change, environmental communication and sustainable development. Lessons learned from interdisciplinary and transdisciplinary research are presented, as well as teaching and classroom methodology for specific countries and disciplines.

Sustainable Energy Education in the Arctic
The Role of Higher Education
Gisele Arruda

Interdisciplinary and Transdisciplinary Failures
Lessons Learned from Cautionary Tales
Edited by Dena Fam and Michael O'Rourke

Environmental Consciousness, Nature and the Philosophy of Education
Ecologizing Education
Michael Bonnett

Theatre Pedagogy in the Era of Climate Crisis
Edited by Conrad Alexandrowicz and David Fancy

For more information about this series, please visit: www.routledge.com/ Research-and-Teaching-in-Environmental-Studies/book-series/RTES

Theatre Pedagogy in the Era of Climate Crisis

Edited by Conrad Alexandrowicz and David Fancy

Routledge
Taylor & Francis Group

earthscan
from Routledge

LONDON AND NEW YORK

First published 2021
by Routledge
2 Park Square, Milton Park, Abingdon, Oxon OX14 4RN

and by Routledge
52 Vanderbilt Avenue, New York, NY 10017

Routledge is an imprint of the Taylor & Francis Group, an informa business

British Library Cataloguing-in-Publication Data
A catalogue record for this book is available from the British Library

Library of Congress Cataloging-in-Publication Data
Names: Fancy, David, author. | Alexandrowicz, Conrad, editor.
Title: Theatre pedagogy in the era of climate crisis / edited by Conrad Alexandrowicz and David Fancy.
Description: Abingdon, Oxon ; New York : Routledge, 2021. | Series: Research and teaching in environmental studies | Includes bibliographical references and index.
Identifiers: LCCN 2020049377 (print) | LCCN 2020049378 (ebook) | ISBN 9780367541545 (hardback) | ISBN 9781003087823 (ebook)
Subjects: LCSH: Theater–Study and teaching. | Climate change. | Theater–Political aspects. | Theater and society.
Classification: LCC PN2075 .T546 2021 (print) | LCC PN2075 (ebook) | DDC 792.02/87–dc23
LC record available at https://lccn.loc.gov/2020049377
LC ebook record available at https://lccn.loc.gov/2020049378

ISBN: 978-0-367-54154-5 (hbk)
ISBN: 978-0-367-76136-3 (pbk)
ISBN: 978-1-003-08782-3 (ebk)

Typeset in Baskerville
by SPi Global, India

Contents

Contributors

Conrad Alexandrowicz, MFA, is an associate professor in the Department of Theatre at the University of Victoria, where he teaches movement for actors. Over a decades-long career in performance he migrated from dance to theatre and has been a dancer, choreographer, writer of texts for dance, playwright, actor, director and producer. He created over fifty dance- and physical theatre works, many of which were presented across Canada, and internationally. His writing has been published in *Theatre, Dance and Performance Training, Studies in Theatre and Performance* and *Theatre Topics*. His first book, *Acting Queer: Gender Dissidence and the Subversion of Realism*, was published by Palgrave in February 2020.

Mary Anderson is the chair of the Department of Theatre and Dance at Wayne State University. She is interested in heuristic processes, the convolutions of remembering through writing and objects in performance. Her articles have appeared in *Performance Matters; About Performance; Body, Space & Technology; Canadian Journal of Practice-based Research in Theatre; Theatre, Dance & Performance Training; Teaching Artist Journal; Research in Drama Education; Journal of Dance Education; International Journal of Education & the Arts; and Arts Education Policy Review*.

Gloria Akayi Asoloko is a Nigerian scholar and development worker with a Master of Arts Degree in Development Communication. She is the co-founder of Framework Advocacy and Development Initiative (FAD Initiative), a non-profit organization aimed at community development in Nigeria. She is a consultant with RNW Media, The Netherlands, where she is involved in developmental projects. Her play, "Who Knows Amanda," won the Society of Nigerian Theatre Artist (SONTA) drama prize and the African Writers Award for Drama. She is currently a staff member with the Institute of Strategic and Development Communication (ISDEVCOM), Nasarawa State University, Nigeria.

Lara Aysal is a climate justice and human rights activist, performance artist and facilitator of community-oriented projects. Her work mainly focuses on migration, ethnic minority conflict and climate crisis. She is one of the co-founders of AA+A Contemporary Performance Research Project and Ray Performance Collective. She gave acting classes at Beykent University. She is interested in the role of theatre to address, organize and take action within climate justice context though decolonizing methodologies. Lara recently completed her artist in residence at Greenpeace Canada and the International Center of Arts for Social Change. She is currently a PhD student in Interdisciplinary Studies at UBC.

Tanja Beer is an ecoscenographer and community artist who is passionate about co-creating social gathering spaces that accentuate the interconnectedness of the more-than-human world. Originally trained as a performance designer and theatre maker, Tanja's work increasingly crosses many disciplines, often collaborating with landscape architects, urban ecologists, horticulturists and placemakers to inspire communication and action on environmental issues. Her most celebrated project is The Living Stage: a global initiative that combines spatial design, horticulture and community engagement to create recyclable, biodegradable, biodiverse and edible event spaces. The Living Stage has been realized in Castlemaine, Cardiff, Glasgow, Armidale, New York and Melbourne, with outcomes of the work exhibited at the Prague Quadrennial of Performance Design and Space and the V&A in London. Tanja is currently a senior lecturer in design (spatial/interior) at Griffith University.

Soji Cole is a theatre creator and scholar. He has a Diploma, B.A., M.A. and PhD degrees, in Theatre Arts, from the University of Ibadan, Nigeria. He has won the African Theatre Association (AfTA) Emerging Scholars' Prize, the International Federation for Theatre Research (IFTR) New Scholars' Prize, the Association of Nigerian Authors' (ANA) Playwriting Prize and the NLNG/Nigeria Prize for Literature (the biggest literature prize in Africa). He was a Fulbright Fellow at Kansas State University, Manhattan, Kansas (2014-15). He taught playwriting and theatre sociology in Nigerian universities, before moving to Canada, where he is currently studying for a second PhD at Brock University.

Derek Davidson teaches playwriting, dramatic literature and theatre history at Appalachian State University. Davidson is also artistic director of In/Visible Theatre (a professional company in Boone, North Carolina), a playwright, director and AEA actor. He has worked as an associate artistic director and coordinator for the Appalachian Festival of Plays and Playwrights at the Barter Theatre in Virginia. Davidson is co-facilitator of the university-wide Climate Stories Collaborative, for which he has conducted several Forum Theatre workshops. His short play "Blackjack," has been performed internationally as part of *Climate Change Theatre Action.*

Katrina Dunn is assistant professor in the University of Manitoba's Department of English, Theatre, Film & Media where she teaches in the Theatre Program. Her scholarly work explores the spatial manifestations of theatre as well as ecocritical theatre. Katrina's long career as a stage director and producer has had considerable impact on the performing arts in western Canada and has been recognized with numerous awards. *Malus fusca* is a North American species of crabapple tree. The singular tree that contributed to this chapter has its roots in Fort Garry in Winnipeg, Manitoba.

David Fancy is professor and chair in the Department of Dramatic Arts, Brock University. He brings his philosophical interest in immanentist thought to performance studies, science and technology studies and critical disability studies. Recent publications include Fancy, David and Hans Skott-Myhre, Eds. *Immanence, Politics and the Aesthetic: Thinking Revolt in the 21st Century.* McGill-Queens University Press, 2019; and Fancy, David and Lillian Manzour Eds. *Teatro de Tres Americas: Antología Norte.* Ediciones Sin Paredes, 2020. Fancy has an extensive practice as a playwright, and director of theatre, opera and circus; he is the editor of a website on the subject of actor training and diversities.

Dennis D. Gupa is a theatre director, educator and dramaturg. His dissertation in applied theatre at University of Victoria focuses on refiguring climate crises, decolonial theatre and the dramaturgy of care. He is a fellow of UVic's Center for Studies in Religion and Society and was a research fellow of the UVic's Center for Asia-Pacific Initiatives. He obtained an MFA in Theatre (Directing) from the University of British Columbia, and an MA in Theatre Arts from the University of the Philippines. He co-authored articles published by Text and Performance Quarterly, Global Performance Studies international and Arts Praxis. Dennis is a Vanier Scholar.

Alexandra (Sasha) Kovacs is an assistant professor at the University of Victoria (Canada) where she teaches and researches Canadian theatre historiography. She is currently completing a monograph concerning the performance history of late Mohawk (Kanien'kehá:ka) Six Nations woman E. Pauline Johnson (Tekahionwake). Kovacs' research is published in *Performance Research, Shakespeare International Yearbook, Canadian Theatre Review* and the collection *Space and Place: Cultural Mapping and the Digital Sphere.* Kovacs is currently a co-investigator on the SSHRC Partnership Development research project *Gatherings: Archival and Oral Histories of Performance.* More information about her research can be found on that project's website: gatheringsparthernship.com.

Rachel Rhoades currently serves as assistant professor of applied theatre at Brock University. She has worked as an applied theatre practitioner, educator and researcher with young people from Grade 1 to the graduate level for 14 years. Her current community-based research interrogates the potential for ensemble devising to serve as a site for cultural sustainability, social resilience and equitable acculturation with refugee and newcomer adults to the Niagara region in Canada. Her doctoral research examined how racialized, socioeconomically under-resourced secondary school-age youth in Toronto conceptualize their current and future roles within contemporary social movements and the larger political sphere through ethnodrama.

Kirsten Sadeghi-Yekta (PhD, University of Manchester) is currently working on her SSHRC Partnership Development Grant and Insight Development Grant on Coast Salish language awakening through the medium of theatre. Her theatre facilitation includes working with children in the Downtown Eastside in Vancouver, young people in Brazilian favelas, women in rural areas of Cambodia, adolescents in Nicaragua and students with special needs in The Netherlands. Kirsten is a faculty member in the Theatre Department at the University of Victoria, BC, Canada.

Caridad Svich is a playwright–lyricist–translator and editor of Cuban-Spanish-Argentine-Croatian descent. She was profiled in the July/August 2009 issue of *American Theatre* magazine. She's received, among others, the 2012 OBIE for Lifetime Achievement, 2012 Edgerton Foundation New Play Award, 2011 ATCA Primus Prize and the National Latino Playwriting Award twice in her career, the 2009 Lee Reynolds Award from the League of Professional Theatre Women, a Harvard University Radcliffe Institute for Advanced Study Bunting fellowship, TCG/Pew National Theatre Artist Grant at INTAR Theatre, NEA/TCG Playwriting Residency at Mark Taper Forum, has been short-listed four times for the PEN USA-West Award in Drama and is featured in the *Oxford Encyclopedia of Latino Literature.*

David Vivian, MFA, ENTC/NTSC, is the scenographer in the Department of Dramatic Arts at Brock University, and Director of Brock's Marilyn I. Walker School of Fine and Performing Arts. Formerly with Concordia University in Montreal, his designs for theatre, film, television and industry have been seen across Canada. His work at Brock has included the set and costume designs for the Mainstage productions since 2004. He teaches theatrical design, production and stagecraft at Brock. David researches marginalized and virtual spaces through visual arts and theatre design, the application of digital technologies to the collection of performance ephemera and regional identity construction and transmission through scenographic practice and research.

Introduction

In the midst of a pandemic and a reckoning on racial injustice

Conrad Alexandrowicz and David Fancy

As we write and assemble this volume engaging futures of theatre education and the climate crisis, the world struggles with the ongoing COVID-19 pandemic, an inundation of the human by the other-than-human that has not been seen (at least in terms of a *viral* inundation) for a hundred years. Like the influenza pandemic of 1918–20, COVID-19 will probably unfold in a series of waves, rather than being contained temporally, so it is unlikely human societies will be able to resume "business as usual" after only a mere number of months. Indeed, there may not be "business as usual" for a long time: in the same way it has gradually dawned on mainstream global public opinion that ongoing and intensifying climate change is in fact best categorized as a crisis, indeed as a global emergency that, as the title of Naomi Klein's 2014 book argues, *"… Changes Everything."* And now with the pandemic—itself a curious and threatening reminder of the agency that "nature" still has over the human, despite millennia of human attempts to eclipse it—the world has changed again, and, it seems to us, everything we do as theatre artists and scholars must change in consequence. "Performance" and "epidemic" are seemingly mutually exclusive terms: traditional performing arts are defined by human co-presence, the very conditions of which have been erased, or severely curtailed, by public health officials undertaking measures to stop or slow the contagion. All over the world, theatre instructors have found themselves forced to adapt their curricula for virtual delivery, an especially doomed project in various sub-disciplines, including performance ("acting," "voice" and "movement"); the pursuits of applied theatre and drama in education; and laboratory work in technical theatre and production.

And yet it continues to be imperative that these crises are understood to inform, intensify and resonate with one other in a variety of immediately apprehensible ways. First: To stop the spread of a highly contagious virus one must shut down a society and therefore curtail its extractive and exploitive globalized economy. A carbon-based economy that is brought to a standstill, while causing financial ruin, will lead to reduced emissions. This has produced a short-lived benefit for the planet's inhabitants, including millions of humans, amid all the lost jobs and emergency government bail-outs. Among other paradoxical effects, consider that while "[a]ir pollution kills an estimated seven million people worldwide every year" ("Air pollution"), these deaths seem to be taken in stride, absorbed as part of what are called "externalities" by economists.[1] At the time of writing, citizens of Delhi can for once see blue sky and breathe something resembling clean air. It is one of the many ironies of this predicament that a comprehensive lockdown of late-capitalist hypermobile societies to stop a virus that primarily produces severe or even fatal respiratory illness—although it has demonstrated its ability to affect many other parts of the body as well—has resulted in the alleviation of conditions (albeit unfolding in a different temporal frame) accepted as routine, which also result in mass fatalities from respiratory illness! For many environmentalists, it is clear that the climate emergency ought to be treated as though it were a pandemic like COVID-19; that is, that we ought to be on a "war footing," transforming everything about the way we live in response not just to what is in evidence around us—although that is dire enough—but to what will arrive if business proceeds as usual and nothing is done. The question remains, however: is the martial language with references to "war," "deployment" and so forth, perhaps not somehow an expression of the deeper conceptual problem informing the emergence of both anthropogenic climate change as well as the COVID-19 pandemic?

Further, one is reminded that the connection between this pandemic—and others in recent memory—and the climate emergency is that both are functions of the overall *environmental* crisis. And that the root of this connection abides in many human societies' instrumentalist, deranged and dysfunctional relationship with everything that is not human, from overfishing and pollution of the oceans with plastics, to the ruination of earth and fresh water with mining waste, to human invasion of every accessible ecosystem. This includes accelerated extractive forces that result in cutting down forests and other habitats that humans ought not to destroy or consume, if only for the selfish reason that these may harbour animals that are reservoirs of pathogens that can kill humans. Or if they can't kill humans directly, they can do so via intermediate hosts, such as has occurred in live animal markets, where trafficked wildlife, domesticated animals and humans are crammed closely together in appalling conditions. Such instrumentalizing practices differ from, but also overlap with, those of industrialized animal farming in the economic north.[2] Of course, it is once again ironic to note how one critical threat is facilitated by another: jet travel, part of the hydrocarbon-fuelled transportation network of globalized capital, is still a central aspect of the climate problem, and it has allowed dangerous microbes to travel in upper middle class comfort all over the globe in a matter of hours.

The third mediatized crisis of the moment, that of global calls for racial justice as part of the ongoing aftermath of 500 years of Euro-American imperialism, is deeply imbricated with both the crisis of the climate and that of the pandemic. We are reminded that this disease, due to the social determinants of health, affects racialized and other vulnerable populations most significantly. Moreover, the history of extractive economic practices that have turned the earth's riches into "resources" to feed capitalist assemblages are themselves deeply dependent on the forced labour of racialized bodies used to gouge those "resources" from the body of the earth (Yusoff 15). From an Indigenous and Indigenous studies perspective, climate change is simply intensified colonialism (Whyte 154), which always already includes "terraforming" that tears away the "'flesh' of human-nonhuman-ecological relationships" (159). Such complex histories and interconnections serve as an important warning that each of us, and each of our students, experiences the intersection of these crises in radically different ways, depending on our/their own positionalities and identifications. Plagues of objectification and extraction; a virulence of racisms; a persistent novel coronavirus: each of these presenting co-morbidities exceeding the temporal frame of an established twentieth-century genealogy relating theatre to disease (Garner 2), projecting us into a twenty-first century, now needing to pursue models of sustainable health in much wider bodies-politic than simply those associated with the (white, liberal, bourgeois) human social body.

Theatre education and the climate crisis

Conrad Alexandrowicz first grappled with the problem of how to transform actor training the face of the climate crisis in the final chapter of *Acting Queer: Gender Dissidence and the Subversion of Realism*, published in 2020. He speculated on the possibilities—as fraught and troubled as they are—of incorporating Indigenous ways of knowing into the actor's work in representing the other-than-human. (This crucial topic is addressed by a number of contributors to this volume.) But the particular idea for this book arose from imagining what we might teach Swedish student activist Greta Thunberg if she were to choose post-secondary theatre education. What approach to acting and theatre-making might we take with someone who excoriates world leaders at the UN summit on climate action, using the phrase "How dare you!" as her motif? (Thunberg). While we doubt that Thunberg would decide to embark on a programme of theatre training as part of her university education—we imagine she is bound for a career in politics (or perhaps virology!?)—the law of averages would suggest that out of the millions of teenagers who have participated and will participate in the world-wide movement of school strikes for climate action, there are many who will decide to do just that. And their pursuit of a theatre education will not position them as any less capable of responding to the climate crisis than undergraduates in other fields. Michael Mikulak, speaking broadly of the role of the arts in the climate crisis, observes that:

> [w]hile it may seem trite to focus on questions of narrative, representation, agency and subjectivity in the face of more 'pressing' material concerns, the environmental crisis is more than a problem for scientists;

it is a problem of narrative, ontology and epistemology. It is as much a failure of the imagination as it is a technological problem, arising from maladapted social and political ecologies that fail to establish healthy and sustainable networks of kinship imaginaries.

(66)

We wonder then, for young people wanting to address the climate crisis: what kind of theatre curriculum would be worthy of both their passions and their fears, their aspirations and their critical capacities and, ultimately, their desire to imagine a different world?

And positing this student of the future is not to suggest that current theatre students' need for guidance in response to the climate emergency is not already in evidence: they are currently creating and presenting work—often devised works configured in the terms of physical theatre—on the climate crisis. They are acting out of the same sense of urgency that is compelling people in all walks of life in countries around the world to wonder: how can we continue with business as usual? How can we *not* address our work—in whatever field we operate—to this subject? Una Chauduri, a pioneer in the area of theatre practice/pedagogy and climate change, has written that

the imaginative and representational work of making art has an enormous role to play in making this unprecedented crisis visible, audible, and felt. ... [W]e argue that theatre is a uniquely powerful site for the kind of thinking called for by the crises of climate change.

(*Research Theatre...* ix)

However, one could equally argue, as many artists and scholars have done, that art never changed anything; that the essential functions of art operate in registers that preclude causality in the zones of *governance*, where decisions with actual material consequences are made.[3] (This debate is taken up and considered by a number of our contributors, as the reader will discover in the following chapters.) Our rejoinder to such objections is that we concur with Chaudhuri's argument not least because we feel we must; because this project has acquired the status of an ethical imperative. If the climate emergency has changed everything, as we argue above, this is how it has changed *us*.

But despite our conviction that the theatre can and should engage with this real-life crisis, many questions remain: what is the theatre of a global emergency? Is there such a thing? If so, how ought it to be practised? And how does one train theatre-makers, inclusively considered, to become its adepts? This book is intended as a kind of provocation to theatre instructors, in their many sub-disciplinary areas, who feel compelled to engage with the project of re-imagining and reconfiguring theatre education in response to the climate crisis. The unprecedented and all-encompassing phenomenon of anthropogenic climate change necessitates the transformation of settled ideas we have about *everything*, including the discourses that support and explicate theatre as an art form, as a mode of philosophical inquiry and as a sphere of pedagogical practice. We propose that such reconfigurations in theatre *pedagogy* imply and call forth transformations in *practice*, given that training inevitably links to and produces it. This book aims to be both a work of reflection and analysis, as well as offering many applicable insights that can make it a practical handbook for instructors.

Concepts, relays, practice

Ecological philosopher Timothy Morton dates our predicament not to the industrial revolution, with its coal-burning steam engines, but to the *agricultural* revolution, which predated the former by 12,000 years, and which, he argues, established the conditions for its later emergence. Those who wrote the Book of Genesis were certainly farmers and herders:

> And God blessed them, and God said unto them, Be fruitful and multiply, and replenish the earth, and subdue it: and have dominion over the fish of the sea, and over the fowl of the air, and over every living thing that moveth upon the earth.
>
> (1:28)

"Agrilogistics"—as Morton calls it—establishes "thin rigid boundaries between human and nonhuman worlds and [reduces] existence to sheer quantity" (*Dark Ecology* 43). Connecting the two revolutions, agricultural and industrial, that are separated by millennia, he claims that *agrilogistics* "is the smoking gun behind the smoking chimneys responsible for the Sixth Mass Extinction Event" (ibid.).

Humans' conceptual split from "Nature," of which the Book of Genesis is constituted, is analogous to the Christian split between "'mind" and "body" that runs through the Epistles of St. Paul, among many other Biblical examples, and finds further philosophical articulation in the work and lineage consolidated by René Descartes. The mind, linked to spirit, struggles with an unruly body whose inevitable functions are identical with bestial Nature, which Man as a generality seeks to subdue. In *We Have Never Been Modern* Bruno Latour analyses the "Great Divide between humans and nonhumans" (97) as the "first dichotomy" (11), from which other sets of divisions have proceeded. "Our intellectual life is out of kilter," he complains. Despite the fact that we are and live in the midst of multiple complex networks, "[e]pistemology, the social sciences, the sciences of texts—all have their privileged vantage points, provided that they remain separate" (5). He asks, "How can we shift … toward collectives of humans and nonhumans?" (77). In *The Politics of Nature* he argues that we must relinquish "Nature" itself in order to save ourselves: "political ecology, at least in its theories, has to let go of nature. Indeed, nature is the chief obstacle that has always hampered the development of public discourse"

(9). Eco-feminist philosopher Val Plumwood, in her classic 1993 text *Feminism and the Mastery of Nature*, observes that the word "nature" ends up partaking of all sorts of absurd conflations, "homogenising in the sweep of 'the rest' things as diverse as seals, waves and rocks, oysters and clouds, forests, viruses and eagles" (70). Western philosophy's profound, consistent and pervasive practice of "radical exclusion ... facilitates the conclusion that there are two quite different sorts of substances or orders of being in the world; for example, mind and body, humans and nature" (ibid.). Our overall—and daunting—conceptual project, both within theatre training as well as beyond it in virtually all enterprises, is thus to move beyond these co-constitutive splits: mind/body = human/nature. If the environment is not *othered* one must take responsibility for it as though it were an extension of one's body; as Morton proposes, "[o]ne solution to the paradoxes of the body is to turn it into the environment itself" (*Dark...* 108).

"The ecological thought"

While many theatre scholars have been working in recent years to re-imagine the art form—both as practice and as a mode of educational inquiry—through the lenses of animal and environmental studies, Una Chaudhuri was certainly one of the first: her work in this subject area, in the form of "The Animal Project," dates back to 2004. (*Animalizing...* 1) Writing about this process of collaborative exploration, dramaturgy and performance-making she articulated the "notion that there may exist a transpersonal and autonomous kind of consciousness, capable of moving and flowing in unforeseen directions, including across species boundaries" (4). More recently Chaudhuri has speculated about performing in relation not merely to nonhuman life-forms, but to that which is not thought to be alive at all:

> How might we practise, think and write theater—and make art—that is aligned with the turbulent planetary present, in which not only humans and other animals but also rivers oceans, coastlines, top soils, groundwater, forests, ice, air, and the atmosphere itself are in a state of constant crisis?
>
> (*Stage Lives...* xiii)

In doing so she summons another notion of the posthuman, that of "vibrant matter," as theorized by Jane Bennett in her eponymous text, in which she speculates that "[i]f matter itself is lively, then not only is the difference between subjects and objects minimized, but the status of the shared materiality of all things is elevated" (3).

These ideas are stimulated and amplified in two more of Morton's notions, that themselves echo each other: first, the "arche-lithic," another invented term that for him captures "a primordial relatedness of humans and nonhumans that has never evaporated" (*Dark Ecology* 63). And second, "the ecological thought," also the title of his 2010 volume, which provides a corresponding component of our argument here. Why the use of the definite article? Morton insists that

> there is a particular kind of thinking that I call *the* ecological thought … Moreover, the *form* of the ecological thought is at least as important as its *content*. It's not simply a matter of *what* you're thinking about. It's also a matter of *how* you think.
>
> (4, emphasis in original)

In other words, he is giving a name to a kind and quality of cognition that is a potential of human consciousness. Morton argues that the effects of the Anthropocene Age have included damage as much to the psychic as to the physical environment; that "they have had an equally damaging effect on thinking itself" (ibid.). It is our very habits of thought that must be remediated, together with the vast and varied kinds of damage that have been done to physical environments. And this process of cognitive transformation may be usefully applied to the paradigms that have governed our values and methods as theatre pedagogues and practitioners. While not exactly toxic, they may require review in light of our drastically changed—and rapidly evolving—circumstances.

Let us assume that as climate-activist theatre pedagogues and artists we propose to follow something we can call "the ecological thought" as a guide for transformation in our work: how might that be articulated theoretically, before we begin to consider practice? One likely route lies across the territory mapped by various eco-feminist thinkers, alluded to above. Val Plumwood's work, among others, explores the unitary source of disparate regimes of oppression, in awareness of the radical possibilities of such identification: "When four tectonic plates of liberation theory—those concerned with the oppressions of gender, race, class and nature—finally come together, the resulting tremors could shake the conceptual structures of oppression to their foundations" (1). Her colleague Ariel Salleh has made the following observation, bracing—perhaps astonishing—in its inclusive sweep:

> Ecofeminism is the only political framework I know of that can spell out the historical links between neoliberal capital, militarism, corporate science, worker alienation, domestic violence, reproductive technologies, sex tourism, child molestation, neocolonialism, Islamophobia, extractivism, nuclear weapons, industrial toxics, land and water grabs, deforestation, genetic engineering, climate change and the myth of modern progress.
>
> (*Ecofeminism* ix)

What we do as instructors and artists is inescapably political, and this fact has only been more explicitly revealed in the context of the climate crisis;

being equipped with an effective set of analytical tools is crucial. It will allow us as instructors to draw connections between the climate crisis and the lived experience of our students, who may routinely suffer various kinds of discrimination, based on disability, sex, race, gender presentation or other factors.

But to bring the focus back to our predicament vis-à-vis humans and their environment: if we hope to heal the profound split in our relationship with the other-than-human, we must employ a model that will allow us to proceed with rigour and insight. Plumwood argues that we must be aware of both our continuity with and difference from other life-forms, which may prevent both sentimentalizing and anthropomorphizing the Other, such that it is neither "alien to and discontinuous from self nor assimilated to or an extension of self" (6).

Climate grief and "Heightened Affect"

The notion that those considered sane may in fact be suffering from psychosis, explored by many authors, such as Gilles Deleuze and Félix Guattari in *Anti-Oedipus: Capitalism and Schizophrenia*, has an intensified significance in the face of the climate crisis. The corporate barons who continue to make enormous profits from the exploitation of fossil fuels seem "mad" indeed, as well as those who either refuse to recognize the crisis they have produced, or who recognize it but refuse to act upon it.[4] And perpetual growth, which has been capitalism's holy writ, the belief by means of which it operates, seems itself a kind of "madness," as it aligns with malignancy.[5] As Thunberg noted in the speech to which we allude above: "We are in the beginning of a mass extinction and all you can talk about is money and fairy tales of eternal economic growth!" While this kind of materially induced and socially inscribed "madness" overlaps with criminality, recent jurisprudence would suggest that those acting in response to these kinds of extractive delusions might well be found not criminally responsible for their actions. We still wonder however, justifiably, if today's oil and gas billionaires are not in fact "criminally insane": what kind of world would we have to live in for them to be either imprisoned or hospitalized? Given this state of affairs, it is no surprise that for those who sit in corporate boardrooms, and for their many enablers in politics, it is environmental activists, in particular those who engage in acts of civil disobedience, such as the Extinction Rebellion protesters or "ecoterrorists" like Earth First, Sea Shepherd or the Earth Liberation Front who are clearly either mad or criminal; but in this emergency such "madness" may be the only hope for survival.

How will we deal with our own "madness" in the teeth of the accelerating crisis? Conrad Alexandrowicz recalls that on the first day of classes in September 2019 the fourth-year actors were describing what they had done over the summer. One of the young men—the most gifted male actor he had ever taught, coincidentally—confessed that over the break he had experienced considerable "climate depression." This was an astonishing thing to hear from a twenty-one-year-old; a statement made in response to an unprecedented condition that is consequently incomprehensible to any other human generation. Similarly, when David Fancy staged an environmentally inflected version of Caryl Churchill's *The Skriker* in 2008, students were invited as part of the training for the production to engage in Grotowskian "Para shamanic" hunting in the further recesses of regional provincial parks (Fancy 103). Part of their response to the work was a deep "climate anxiety" and "climate grief" in response to their own sense of complicity with the mechanisms of extractive and exploitive economic practices. It is commonly understood that the arts have, broadly speaking, "'healing" functions: consider the practices of art, dance and drama therapy; and much applied theatre has tremendous enlivening and affirmative effects in the communities in which it is undertaken. As such, we understand that the kinds of pedagogical practices articulated in this volume, and further teaching and learning strategies that can be developed from them, can reposition human subjectivities in a way that can contribute to overcoming the pathologizing dualities that emphasize both the radical separateness between the "human" and "nonhuman," and the unquestioned supremacy of the former over the latter. In such a model, sustainability becomes less of a potentially reductive economistic articulation of resource allocation, and instead more about "de-centering anthropocentrism in the new complex compound" that Deleuze and Guattari consider to be "nomadic subjectivity" (Braidotti 6). The nomadic figure in this instance is not that which is distanced from community; rather it is one that does not become restricted by limiting, unsustainable notions of the hermetically sealed identity of the bourgeois subject. Instead, this flexible, mutable subjectivity

> brings together ethical, epistemological and political concerns under the cover of a non-unitary vision of the subject. 'Life' privileges assemblages of a heterogeneous kind: animals, insects, machines are as many fields of forces or territories of becoming. The life in me is not only, not even, human.
>
> (Braidotti 6)

These meanings and practices can be useful, inspiring theatre artists to stage futures beyond the anthropocentric, but can also serve the increasingly acute mental health needs of the communities—inclusively considered—that are post-secondary theatre departments.

Realism and its discontents

Moving beyond the anthropocentric, and the various aesthetic, political and psychological salves such a move can engender, is thus a major theme in this volume. To put it another way, invoking the inventions of Timothy Morton, cited above: fundamental to the notion of animating "the ecological thought" in theatre pedagogy is the need for the performer to cast her/ his gaze outward to the nonhuman. A concomitant implication of this is that the often politically and aesthetically restrictive tenets of psychological realism that continue to be central to a significant amount of theatre training are to be challenged. Many of the strategies for teaching to engage life beyond the human are captured by Luce Irigaray's recent questions about the possibilities of explorations of and coexistence with vegetal being: "Can I communicate it, and how, without betraying it, without forgetting it and forgetting myself in such a forgetting: without losing both it and myself? In other words: can I still return among humans, and through what path?" (7). Accordingly, contributors Alexandrowicz, Mary Anderson, Katrina Dunn and Fancy, in differing ways and from disparate perspectives, all grapple with the conceptual and physical tasks associated with perceiving, embodying and representing the other-than-human. This has implications for, and may be applied to, devising work for performance, but it is also a pedagogical end in itself. And the latter, with its implication of ritual and active meditation, and therefore notions of spiritual practice, opens the door to a vast terrain of study: the link between art and religion, between creative expression and notions of the divine, as troubled—and troubling—as these may be. We are put in mind, considering how "playing the nonhuman" might manifest itself in the acting classroom, of the age-old Indigenous traditions of shamanism. Simply approaching these, let alone attempting to partake of them, is inevitably problematic, as we acknowledge in an endnote to this introduction. Is it time for us as deracinated, mostly secular people, many of us living in settler societies, to invent *our own* versions of shamanism?

Alexandrowicz, Lara Aysal, Derek Davidson, Dennis Gupa, Rachel Rhoades and Kirsten Sadeghi-Yekta all address topics concerning devising, both in communities of theatre students, and in those outside post-secondary institutions. For those of us old enough to remember, this was called "collective creation" in the 1970s, the decade to which one can date the origins of this multi-faceted practice, at least in the "West." Climate-conscious performance objects will likely escape the frame of realism, and will be activist in disposition, thus summoning both the tradition of "political" theatre, and the vital pedagogical area known as applied theatre. The crucially important work of Augusto Boal is considered by various writers in this collection: Boal's "Forum Theatre," in these discussions, may now be applied to a problem with truly global reach.

Sasha Kovacs, meanwhile, considers the obstacles to the production of climate-activist performance that abide in the production regimes of most theatre departments. Many of these are aligned with the North American regional theatre model, a structure, with its lists of subscribers usually seeking a "good night out at the theatre," that fits neatly within the overall schemes of the corporatized, neoliberal university.

Just as we must *think* differently before we can *act* differently—in any number of that word's meanings—so we may find ourselves having to use language in unprecedented ways: Playwright Caridad Svich, in conversation with the co-editors, speaks of the new conceptions of "dramaturgy" she urges her students to embrace, including "plant-urgy" and "water-urgy," wherein the very structures of story-telling may be considered metaphors of other-than-human structures; perhaps another instance of *the ecological thought?*

Playwrights/scholars Soji Cole and Gloria Asokolo offer a troubling perspective on the climate crisis from Nigeria, a major player in the global South, where rampant fossil fuel extractivism, widespread poverty and entrenched government corruption form a deadly compound. They reflect on the potential of the climate-activist drama *Wake Up Everyone*, by Greg Mbajiorgu, to shift public opinion on climate change.

Finally, designers and scholars Tanja Beer and David Vivian in conversation with co-editor David Fancy consider the materiality of theatre design and production as a fundamental and immediately apprehensible component of this conversation: how can set, lighting and costume design be undertaken in a sustainable manner? How do the concerns they evoke resonate with other aspects of theatre production, such as the design of theatre buildings themselves, as well as the practices of the audiences who attend them?

Extended and Indigenous temporalities

Theatre always retains a culturally bound set of expectations around the temporality of a specific moment of performance. A "long" three-hour production might seem like a very "short" performance in a cultural milieu where multi-day or even longer performance cycles are the established norm. If time-space compression is a constant in theme in sociological accounts of postmodern societies (and the even more recent articulations of capital and accelerationism), then as Jeanne Tiehan suggests in "Climate Change and the Inescapable Present," we accelerated late-capitalist humans (particularly in the economic North) are currently in a temporal paradox: "we are confronted with an experience humans have never faced, and we remain slow to respond to it with the urgency climate scientists advocate" (123). To riff off Coleridge's famous statement about the simultaneous ubiquity and yet inaccessibility of water in *Rime of the Ancient Mariner*. "Time, time everywhere, nor any moment to act." Part of climate-conscious theatre training in post-secondary education involves changing frames regarding temporality: critical engagement, and the action that may proceed from it, must be accelerated, while the objects of this engagement operate in vastly expanded temporal frames; in the vast slowness of geological time. Fancy addresses this conundrum in his meditation on "geoartistry," while Aysal and Gupa, in probing Indigenous perspectives, acknowledge notions of time that are very different from those of the globalizing West. Elsewhere than in this volume, Kyle Whyte, in his essay on Indigenizing futures and decolonizing the Anthropocene, draws attention to culturally informed perspectives on climate and temporality, abiding both in ritual and the "performances" of everyday life:

> Indigenous climate change studies perform futurities that Indigenous persons can build on in generations to come. That is, our actions today are cyclical performances; they are guided by our reflection on our ancestors' *and* on our desire to be good ancestors ourselves to future generations.
>
> (160)[6]

How to use this book

We've stayed within some fairly traditional organizational modes with this volume, with an introduction and various sub-sections organized along sub-disciplinary lines (performance, production, applied theatre and drama in education, theatre/performance studies), in accordance with the way most post-secondary theatre departments are configured. But this is not to say, as the investments in theatre and performance, in science and sociology, and in theory and philosophy in this introduction suggest, that there is not great interdisciplinary and *intra*-disciplinary potential here. Nonetheless, recognizing the professional designations and disciplinary arrangements within curricular areas will, we hope, provide instructors with the opportunity to vector quickly into their major field of interest, while at the same time not foreclosing exploratory reading through lateral areas of interest. This kind of open-ended and proliferative reading can also help engender the types of deliberations about how to plan for the integration of pedagogies and curricula that are responsive to the climate crisis across a department, or in networks of departments across campuses. It is our conviction that practice and theorization, or creativity and analysis, rather than being mutually exclusive binaristic pairs, are, following Deleuze and Foucault, best understood as mutually informing relays:

> Practice is a set of relays from one theoretical point to another, and theory is a relay from one practice to another. No theory can develop without eventually encountering a wall, and practice is necessary for piercing this wall.
>
> (Deleuze 2004a, 207)

As a result, we invite readers to recognize a whole continuum of different types of intervention the chapters can engender, from appraisals and audits of production practices, to reconceptualizations of mimesis across a curriculum, to specific strategies for engaging youth in applied theatre settings and in performance studios. With a view to guiding each of the chapters towards pedagogical practices, each chapter finishes with a bulleted section called "Practically speaking…"

Rather than finishing with a "conclusion" or even an "afterword," we instead launch our readers into the "beyond" of the volume via a collectively authored manifesto that includes a summary of major insights from all of the contributors' work. In a tip of the hat to one of the patron gods of theatre in the Greek tradition, Deleuze affirms that the book, with all of its realities as a phenomenon deeply invested in Western modes of knowing, can nonetheless be a "Dionysian machine" (2004b, 263) that exceeds the ability of the human author and reader alike to impose a reading on it. As such, we invite our readers to engage with these chapters—and the concepts, languages and practices they harness—with a view to intensifying and leveraging all of their own connections, networks and professional engagements to making theatre and higher education more responsive to the climate crisis.

Notes

1 See International Monetary Fund. https://www.imf.org/external/pubs/ft/fandd/2010/12/basics.htm (accessed 15 May 2020).
2 See Louise Boyle, "The Independent calls for tighter restrictions on wildlife trade and markets." *The Independent*. https://www.independent.co.uk/environment/wildlife-trade-market-ban-restrictions-china-wuhan-a9459796.html (accessed 13 April 2020).
3 For example, consider Hans-Thies Lehmann's scepticism regarding theatre's "political" potentials in his highly influential *Postdramatic Theatre*. Translated and with an Introduction by Karen Jürs-Munby, Routledge, 2006, pp. 174–186.
4 See Bruno Latour's *Facing Gaia: Eight Lectures on the New Climatic Regime* for a stimulating articulation of these overlapping forms of denial.
5 "Metastasizing Capital: the Logic of Unbridled Growth," was the title of a symposium at Brock University in 2006 organized around philosopher Antonio Negri's first visit to North America.
6 We recognize that reference to Indigenous epistemologies is itself always already at risk of being itself exploitive in terms of intellectual "capital," particularly as this has become in vogue in higher education over the past few years. On this subject Whyte notes, in a fashion that is equally applicable to theatre and performance studies, and the subject of this volume, "[o]f course, many Indigenous persons are understandably concerned that climate scientists will intentionally or naively clamour around Indigenous communities to exploit the information Indigenous knowledges might possess that could fill in gaps in climate science research" (159).

Works Cited

Air Pollution. World Health Organization (WHO), 2020. https://www.who.int/health-topics/air-pollution#tab=tab_1 (accessed 16 July 2020).
Alexandrowicz, Conrad. *Acting Queer: Gender Dissidence and the Subversion of Realism.* Palgrave, 2020.
Bennett, Jane. *Vibrant Matter: A Political Ecology of Things.* Duke UP, 2010.
Braidotti, R. *The Ethics of Becoming Imperceptible*, 2008. http://deleuze.tausendplateaus.de/wp-content/uploads/2008/01/trent-final.pdf (accessed 8 July 2020).
Chaudhuri, Una. *The Stage Lives of Animals: Zooësis and Performance.* Routledge, 2017.

Chaudhuri, Una and Shonni Enelow. "Animalizing Performance, Becoming-Theatre: Inside Zooësis with the Animal Project at NYU." *Theatre Topics*, vol. 16, no. 1, 2006, pp. 1–17.

———. *Research Theatre, Climate Change, and the Ecocide Project: A Casebook*. Palgrave, 2014.

Coleridge, Samuel Taylor. *Rime of the Ancient Mariner*. 2020. https://www.poetry-foundation.org/poems/43997/the-rime-of-the-ancient-mariner-text-of-1834 (accessed 25 August 2020).

Deleuze, Gilles. "Intellectuals and Power: A Conversation between Michel Foucault and Gilles Deleuze." In *Desert Island and Other Texts, 1953–1974*, edited by David Lapoujade, translated by Michael Taomina. Semiotext(e), 2004a, pp. 206–213.

———. *Logic of Sense*. Translated by Mark Lester with Charles Stivale, edited by Constantin V. Boundas. Continuum, 2004b.

Fancy, David. "Sustainability, Immanence, and the Monstrous in Caryl Churchill's *The Skriker*." In *Sustainability and the Globalized Classroom*, edited by A. Shannon and Richard Moore. Sense Publishers, 2015. pp. 95–106.

Garner, Stanton B. "Artaud, Germ Theory, and the Theatre of Contagion." *Theatre Journal*, vol. 58, no. 1, 2006, pp. 1–14.

Irigaray, Luce and Michael Marder. *Through Vegetal Being: Two Philosophical Perspectives*. Columbia UP, 2016.

Klein, Naomi. *This Changes Everything*. Vintage Canada, 2014.

Latour, Bruno. *We Have Never Been Modern*. Translated by Catherine Porter, Harvard UP, 1993.

———. *The Politics of Nature: How to Bring the Sciences into Democracy*. Translated by Catherine Porter, Harvard UP, 2004.

Mikulak, Michael. "The Rhizomatics of Domination: from Darwin to Biotechnology." In *An [Un]Likely Alliance: Thinking Environment[s] with Deleuze|Guattari*, edited by B. Herzogenrath. Cambridge Scholars Press, pp. 66–83, 2008.

Morton, Timothy. *The Ecological Thought*. Harvard UP, 2010.

———. *Dark Ecology: for a Logic of Future Coexistence*. Columbia UP, 2016.

Plumwood, Val. *Feminism and the Mastery of Nature*. Routledge, 1993.

Salleh, Ariel. "Foreword." In *Ecofeminism*, edited by Maria Mies and Vandana Shiva, Zed Books, 2014.

Thunberg, Greta. "Greta Thunberg to World Leaders: 'How Dare You? You Have Stolen My Dreams and My Childhood." *YouTube*, Uploaded by Guardian Online, 23 September 2019, https://www.youtube.com/watch?v=TMrtLsQbaok (accessed 9 July 2020).

Tiehin, Jeanne. "Climate Change and the Inescapable Present." *Performance Philosophy*, vol. 4, no, 1, 2018, pp. 123–138.

Whyte, Kyle. "Indigenous Climate Change Studies: Indigenous Futures, Decolonizing the Anthropocene." *English Language Notes*, vol. 55, no. 1–2, Fall 2017, pp. 154–162.

Yusoff, Kathryn. *A Billion Black Anthropocenes or None*. University of Minnesota Press, 2018.

Part 1

Applied theatre/drama in education

1 Nurturing hopeful agency

Applied theatre pedagogy in collaboration with social movements

Rachel Rhoades

This chapter interrogates the potential of collaboration between applied theatre courses in higher education settings and social movements as a means to nurture hopeful agency in university students, who are often overwhelmed by issues such as climate catastrophe and global capitalism. I share the pedagogical praxis that I enacted in my doctoral research study, *Youth Artists for Justice: Examining Participation in Social Movements and Envisioning Futures through Applied Theatre*, and in a course I currently teach entitled "Social Issues Theatre for Community Engagement." The Youth Artists for Justice study examined how twelve racialized, socio-economically under-resourced youth in later years of secondary school envision their futures, and the possible forms of political participation they may undertake, given developing and inescapable environmental crises. In the "Social Issues…" course, third- and fourth-year students are developing skills and applied opportunities to facilitate applied theatre with marginalized groups within various contexts by applying theoretical and practical knowledge. A goal of both the study and the course is to garner within young people a sense of hope and capacity to conceptualize and enact their political agency.

This chapter contributes to scholarly discourse on the potential for linking drama pedagogy with youth participation in social movements as a mechanism for nurturing critical hope and collective agency in the face of widespread despair when confronting the existential threat of climate catastrophe. I examine Collective Forum Theatre pedagogy as a starting point to support youth in further developing theories of change, envisioning their own potential political roles and entering into participation on their own terms. I use the term "collective" in describing a particular version of Forum Theatre that I apply with students, in which the focus is not on a sole protagonist but rather on an accumulation of protagonists, that models the need for solidarity and diverse strategies in undertaking resistance. I argue that applied theatre pedagogy has the potential to deepen analysis that honours and applies counter-hegemonic knowledges, that embraces and mobilizes emotional expression, that disrupts power, that enriches collective agency and that creates opportunities to perform a public performance with political efficacy. All of these attributes support not only the development of political agency and critical hope within and among university students, but also enhance the process of strategy development and solidarity-building within social movements. I illustrate and interrogate pedagogical techniques that enable classroom communities to engage in building creative, collective, mutual learning, while also enriching the affective desire for expression and political influence. Collective Forum Theatre pedagogy, in collaboration with social movements through community engagement, may serve as an opportunity for young people to take on the roles of public pedagogues and movement strategists.

Context: neoliberal expectations imposed upon youth

University courses in applied theatre can provide an opportunity for political education and direct participation that is out of reach for many students in the neoliberal climate, whose discourse dictates that they focus on their futures as contributors to and potential beneficiaries of economic growth. In addition, from a developmental perspective, many adults view youth as emerging political subjects in need of training and empowerment, rather than as capable agents of change.[1] In a study on various youth activist groups and perspectives of youth in Canada, Kennelly reports on the pressures imposed upon students to be "good citizens" and "to placate youth activism into acceptable forms of liberal individualism" (9) by grooming students into neoliberal-minded workers strengthening the national economy. Gordon critiques the messaging students receive, namely that they are mere "citizens-in-the-making …where young people's political participation can be imagined, but only in terms of their adult eventuality" (8). Lesko expresses scorn in her scholarship for the ways in which adults coerce youth to develop "human capital [for the purpose of] producing useful citizens, [therefore] abandoning them to the global economy" (182). For critical youth scholars and critical educators of young adults, these obstacles translate into the need to conduct research and to enact pedagogy that attend to the political intentions, choices and activities of youth, and to popularize discourse on their commitment and involvement.[2]

Slater and Briggs state how the market mindset driven by expanded opportunities for profit has created an "educational marketplace" in which students become "units of productive [and] consumptive potential" (443). Belay, a seventeen-year-old second generation Somali Canadian male participant in my research study, commented on the neoliberal individualist messaging he receives in educational settings:

> I would say it's more kind of like an ideology. When you're a little kid, when you're in school, one of the first things teachers ask you [is] "What are you good at?" They want to know what you specialize- … There's always two sides of what you're good at and what's needed. So, I guess the whole thing would be society's view of work on yourself, like a selfish view, a general selfish view.
>
> (June 21, 2017)

Belay gestures to a desire to locate oneself not in terms of contributions to the market sphere, "what you're good at," but rather in relation to the greater push for "what is needed" in the world. Critical youth studies scholar Kennelly warns of the discourses of developmentalism and youth citizenship, which often undergird the goals of schooling, that are "about identity and characteristics seen as valuable by the state ... [and] function as form[s] of governmentality" (25). Many of the youth participants felt out of place in school settings, marked as they are by neoliberal values. Even in social environments, surrounded by friends, young people are still taught to capitalize on social relationships, or disregard them if they do not help one develop as a marketable individual. This neoliberal ideology, manifested in the school settings where youth spend the majority of their waking hours, only increases a sense of displacement, of alienation from the present moment and from relational ties. While youth have the capacity for resilience, the isolation and focus on competition espoused in schools can place a strain on them, particularly when they are already overwhelmed when imagining the near and distant future of climate crises.

Reclaiming collectivity and democratic imagination through applied theatre

Intimate relational ties and regard for the power of collectivity are both key components of drama pedagogy. In particular, applied theatre is characterized by ensemble community-building and collaborative participation as fundamental in the process of cultural production, with aims to effect progressive change in marginalized communities. Jonothan Neelands discusses the unique and crucial social nature of ensemble-building as community development within applied theatre, which he directly connects to experiential democratization or "the practice of freedom" (*Taming the Political...* 316). Neelands claims

> [that] participatory forms of theatre have the greatest democratising potential. ...Active participation through dialogue and interaction is necessary in both the making and the experiencing of theatre – a participatory theatre then pre-supposes the engagement of social actors who have the possibility of also being artistic actors.
> (*Democratic and Participatory Theatre...* 36)

Berardi describes this historical era as one characterized by "social precarity," which he defines as "the condition of existential loneliness coupled with all-pervading productivity" (146). I argue here for an applied theatre pedagogy that incorporates collaboration with climate justice movements as a means of enhancing hope in students through hands-on contributions gleaned from collective creative action. Thus, as Neelands argues, the student ensemble may enact democratic participation through cultural production, in this case, in support of climate justice (46).

Applied theatre has particular resonance now, as critical youth studies and drama pedagogy scholar Kathleen Gallagher argues:

> [G]lobal, political neo-capitalist interest ... has insidiously entered into the level of local public discourse, rendering very difficult imagined possibilities outside the logic of capitalism. Theatre is one distinctive place ... where we might think the present otherwise, even though discursive and material forces of neoliberalism continue to threaten its radical possibilities.
>
> (55)

Applied theatre courses can establish a site in which to explore such "imagined possibilities" through creative, critical praxis, that is, a cyclical process of information-gathering, devising, reflection and performative action. Applied theatre in particular can model collectivity in devising new modes of existence, with its emphasis on the power of collective problem-solving and the creative enactment of critical thinking. I designed the "Youth Artists for Justice" drama programme and my "Social Issues Theatre for Community Engagement" course with the intention of providing a space for strategizing and positing ways of participating in politics and impacting society through applied theatre. In the following section, I share my critical pedagogy praxis, that I argue may support drama students in gaining hope through deepening their knowledge of effective youth activism and cultural production as a social movement strategy.

Youth artists at the helm of social movements: galvanizing hope through peer exemplars

During the "Youth Artists for Justice" study, and in my applied theatre course, I introduced students to many strategies, philosophies, struggles and accomplishments of social movements that engaged in various forms of resistance, direct action and cultural production;[3] in particular, I shared youth-led arts activist exemplars. By offering case studies of youth activism that had significant impact through the arts, I aimed to provide the participants with a sense of belonging and momentum, of hope and inspiration. Sharing these examples also fuelled dialogue on the youths' creative, pedagogical and political intentions, as well as the potential principles and strategies they might implement as theatre-makers involved in struggles for justice. I facilitated this discussion to enable the youth to participate on creative, practical, ethical and theoretical levels when considering applied theatre as a mode of effecting change in social movements.

I provided models for youth participants of contemporary versions of youth-led cultural production that explicitly name and challenge social inequities, promote new social relations and combat hegemonic domination. My aim was to provide education on social movements that youth often do not access through formal educational channels, and to support them in developing and presenting their own ideas about creating change through applied theatre, with the overall goal of increasing their sense of hope for the future in terms of their own political agency.

I had the honour of working as a research assistant with Kathleen Gallagher on a project in which we collaborated with drama education scholars and practitioners across five countries, including Urvashi Sahni, the founder of The Prerna School for lowest-caste girls in Lucknow India. Sahni utilizes drama pedagogy in conjunction with critical feminist pedagogy, and the mission of the school, in part, is to "raise their feminist consciousness and to help them emerge as emancipated women with a perception of themselves as equal persons having the right to equal participation in society" (Sahni 60). Veerangna is the direct-action drama group that Prerna students engage with as a means of participating in social change through initiatives such as raising awareness about and reducing domestic violence. Through a combination of public performance, organizing rallies, creating a help phone line, collecting petition signatures declaring a commitment to ending domestic violence and engaging the police force, the young women take on the roles of public pedagogues and social justice activists, anchored in applied theatre practice. I shared one of the Veerangna street performance videos with youth and students to bolster solidarity and provide an inspiring example of applied theatre as social resistance.

Xiuhtezcatl Martinez, an Indigenous climate justice artist-activist, is the youth director of the global Earth Guardians organization, made up of thousands of youth demanding environmental sustainability through various forms of protest from lawsuits to marches. Since his childhood Martinez has protested through speech and music, such as with the hip-hop song, *What the Frack?*, which helped mobilize the political movement to successfully ban fracking in Colorado and other areas in the United States. At the age of fifteen he spoke as a representative of civil society at the General Assembly of the United Nations before world leaders from 193 countries. I showed youth Martinez's speech from the Youth Leadership Bioneers Conference, in which he proclaims:

> What better time to be born than now? Because this generation gets to rewrite history, gets to leave its mark on this Earth, because this generation will be known as the people on the planet that brought forth a healthy, just, sustainable world for every generation to come, because this generation of people get to create the rebirth, get to co-create and re-create this new world of sustainability, of justice, of equity for all people on Earth. We are the generation of change. We are generation RISE!

Exposure to youth artist activism is one important step towards nurturing hope and a sense of political agency in youth, who often feel distanced from participation and the potential for leadership within social movements.

Here I offer a number of other case studies involving recent youth-led resistance actions on a global scale that may inform and inspire students. There are a variety of lessons to be gleaned from each of the cases, including the constant shift of tactics required as political conditions change such as in the case of Egypt (Abdalla 2016); the need to identify which gatekeepers hold the greatest power in relation to the goals of the struggle as in the case of Shannen Koostachin in Attawapiskat (Angus 2015); potential strategies to galvanize peers as in the case of the DREAMERS (Gamber-Thompson and Zimmerman 2016); and the vitality of cultural production as a medium for instigating change as in the cases of Venezuela (Jaramillo 2015) and Palestine (Desai 2015).

Experiential education in youth artists for justice: instilling hope and agency

Interest in the Youth Artists for Justice programme stemmed in great part from young people's desire to learn more about social movements and about different forms of participation, including activism through the arts. One of the examples I used is the work of Black Lives Matter-Toronto (BLM-TO). Among the quotes I shared is one from lead organizer Syrus Marcus Ware: "BLM TO is actively blurring the lines between direct activism and artistic practice. We are rooted in freedom fighter artists worldwide and throughout the centuries" (Art Creates Change, Ontario College of Art and Design, October 26, 2016). I introduced the youth participants to the BLM-TO action at Pride Toronto, in which they held a sit-in and used carnival-style presentation with a float covered in images of historic Black transwomen activists, and invited them to a BLM-TO protest event. I also invited the youth to join the Rhythms of Resistance (ROR) band that I played with as part of the National Day of Action Against Islamophobia and White Supremacy organized by BLM-TO. ROR consists of many seasoned activists, and we have established protocols to maximize our safety and cohesion during protest events. While there is undeniable risk associated with such events, I felt that by engaging in the act of protesting the participants could gain a sense of solidarity that they could counteract the anxiety and feeling of powerlessness they expressed throughout the pre-programme interviews. Ideally, the youth who attended would bring more nuanced perspectives to the complex analysis of youth roles in social resistance. In deciding upon which protest to which to invite the youth, I felt confident that a rally organized by Black Lives Matter-Toronto, a well-established and well-connected chapter, would likely offer a positive experience for them.

I trusted from my experience that these organizers would model resistance and solidarity with skill and passion. I anticipated that the protest might bring the youth more hope and also allow them to open their imaginations and heighten their ambitions as they developed relationships and theatrical work with their peers about their roles as change agents. Aziz Choudry states in the *Learning Activism* that "[l]earning is social.

...[P]eople's everyday practices in struggles contribute to constructing alternative forms of knowledge ... [and] for those who want to change the world, this kind of knowledge and learning offers important tools for political praxis" (81). When envisioning the ideal scenario for youth participants to engage in activism through attendance at the protest, I recalled how Harney and Moten describe the process of creating emancipatory scholarship, in their case with a community of Black thinkers. When I first read their depiction of the "cacophony and noise [that] tells us that there is a wild beyond to the structures we inhabit and that inhabit us" (7), it reminded me of the exhilaration and sense of possibility and purpose that I have experienced through activism.

The youth participants who attended described observing a particular set of social relations in which a powerful sense of safety and possibility emerged. This affective response emerged from being surrounded by a community of diverse peoples expressing concern, rage and love for those harmed by systemic injustice. The youth experienced an ideal moment in the process of building trusting relationships between groups in the form of the protest and the ensuing march, where all of them played shakers as musical instruments along with the band.

In his post-programme interview, Mike, a Black twenty-year-old shelter-dwelling male, reflected on his protest experience:

> We, first of all, get the chance to let people know that, "hey, again, there are people standing with you." I think there were over a thousand people that were there, letting an entire group of people know that it doesn't matter what you choose to believe, it doesn't matter what religion you are, what your gender is, what you choose to love, there are people here and there are people who will be here to fight for you.
>
> (interview, June 8, 2017)

Mike articulated the concrete power of solidarity in silencing bigotry, and the sense of safety and belonging to a voluminous community of care and acceptance at the protest. His reflection brings to mind Paulo Freire's[4] critique of social movements in the United States in 1996. Freire argued that there was a lack of cohesion between groups struggling against a common set of power structures that subjugated them all, however different the contexts. Freire spoke of "the necessary solidarity which people who have the same dreams or similar political dreams have to have among themselves in order to struggle" (62–63). He went on to discuss the need to admit and understand differences between the various groups engaged in struggle, and that solidarity needs constant tending. Jenaya Khan, a BLM-TO organizer, describes the continuing struggle to develop meaningful relationships across identity groups, particularly in Canada:

> Here, for example, because black people on their own could never reach critical mass—we don't have the population for it—we have to

be very deeply invested in solidarity movements. [...] Those connections are things that take a lot of trust building and time.

(Gomes and Khan 54)

In the following section, I articulate my particular approach to Forum
Theatre as enacted in the "Social Issues Theatre for Community Engagement" course that attends to the concern Khan expresses above, particularly concerning the need for solidarity in social movement organizing.

Interweaving political education and applied theatre training in a university setting

Before delving into the collective forum theatre pedagogy/social movement strategy, it is important to frame the current context for youth activist
participation and the potential benefits of a university setting in developing vital knowledge and skills for effecting change. Universities may provide an ideal environment in which to nurture intergenerational activism
that honours the contributions of youth, while also maximizing intellectual
and strategic transfer of knowledge from professors and activist leaders
with whom the class collaborates. Braxton posits the need for young people in current times to build "autonomous spaces to utilize their creativity and create their own demands and tactics" (37), while also acting at
times in coalition with adults in order to maximize their capacity as political participants. Sekou Franklin calls for a "bi-directional character" (256)
to intergenerational collaboration, in which adults contribute resources,
links to established organizations and political education. In turn, youth
inform adults of the issues impacting their communities, communicate
their demands for change and share their ideas for achieving more equitable conditions.

Applied theatre is a site that allows people to engage in collaboratively creating and modelling possibilities that counter the status quo. Prendergast
and Saxton describe applied theatre as "a collective approach to creating
theatre pieces in which the makers themselves become aware and capable
of change" (11), one that "works overtly either to reassert or undermine
sociopolitical norms, as its intent is to reveal more clearly the way the world
is working" (8). In my doctoral research and applied theatre course, the
discipline became a platform for "cultural activism" and "political activism
through performance-making" (Nicholson 11) with the youth participants,
as they envisioned and enacted forms of resistance to the inequities they
observed in their own lives, and those experienced by other marginalized
groups.

Collective Forum Theatre maximizes the experiential expertise of various community members and creates opportunities for people from diverse
backgrounds and distinct contexts to share their knowledges regarding
political engagement. Choudry describes the critical informal learning
that occurs in social movement participation and advocates for diverse

forms of collective approaches to effecting change and sustaining healthy, equitable social relations amongst organizers and activists. These strategies include "[s]imulation and role-playing games [that] can engage people about power relations and systems that stratify people" (92). Choudry uses the example of Forum Theatre, a process which he describes as "conscious intervention, as a rehearsal for social action rooted in a collective analysis of shared problems" (93).

Below I offer the example of devising a Collective Forum Theatre scene with university students on the topic of climate change, in which the participants embody their learning around social movement strategies, and the need for solidarity across groups as fundamental to formulating steps to effect change and achieve justice. Prior to the Forum Theatre exercises, students engaged in weekly readings and class discussions, wherein they examined the principles of social movement in tandem with applied theatre case studies.[5]

The students collectively decided to focus on climate justice as the topic of our Forum Theatre exercise. In one student's words: "Obviously, if you think about all the social issues that we have and that are possible to explore, they're all very important, but I feel like none of them will matter if there's no Earth" (class discussion, March 2, 2020). We decided that the antagonist should be employed in the upper-level administration of a dairy product corporation, given the personal experiences and political knowledge of a few students. What started as a single student—wracked with concern, despair and rage—confronting a corporate leader who was out having a smoke break, turned into a coalition of people planning a diversity of strategies to effect change. These characters included: a dairy farmer, an employee at a local branch of the corporation, a cashier at a grocery store, a journalist for the regional paper and a young activist, along with the original university student. Students applied the concepts analysed in readings and class discussions throughout the Collective Forum Theatre activity, such as "radical flanking," which McBay articulates in relation to the successful historical movements, such as the direct action undertaken by the British Columbia Wimmin's Fire Brigade's against a violent pornography chain in 1982, and the militancy of the Deacons for Defense in the US South prior to and during the Civil Rights Movement. Radical flanking is the idea that "militant action makes a moderate position (and the possibility of compromise) much more appealing to those in power, and it makes formerly risky action appear more moderate" (85), thus pushing more radical ideas further along The Overton Window spectrum of unthinkability towards popularity.[6] The Overton Window is an analytical tool used in political science to measure a spectrum of public opinion in regards to social movement tactics/organizations (Unthinkable-Radical-Acceptable-Sensible-Popular-Policy). In relation to Collective Forum Theatre, McBay's provocative queries, such as, "can these tactics maximize our political force, and direct that force intelligently?" (98), are helpful guideposts for rehearsing resistance.

What can collective forum theatre contribute to climate justice movements?

The unique feature of Collective Forum Theatre is that it involves an accumulation of protagonists with diverse experiential knowledges, such as a social movement organizer, a racialized critical theory educator, a minimum-wage worker and an alternative media producer. Usually in Forum Theatre there is one main protagonist in a scene, which repeats itself with a new spect-actor[7] per iteration, attempting a distinct strategy for combatting the oppressive antagonist figure. In this form, the responsibility is not solely on the individual. One student reflected on the relevance of this detail in connection to how people envision effecting large-scale changes:

> I liked how instead of talking about- well, if we just recycle, if we turn out the lights for one hour a day, as individuals the problem will go away, and instead, we actually talked about what the systemic issues were and the corporations that are feeding into this problem and how we can organize against these corporations so that our voices are heard, rather than putting the onus on the individual, because it's not an individual thing.
>
> (classroom discussion)

Brucato describes our current social climate as overwhelmingly individualistic, an effect of pervasive neoliberal ideology, to the point that "citizens [are] isolated from one another and thus underprepared to act collectively" (46). According to the students, Collective Forum Theatre not only provides a platform for people to mobilize and enact communal social agency, but also encourages listening and open-mindedness:

> I think it's really valuable, too, to have people playing different perspectives than what they may personally have, 'cause that's just helpful in the general world, just challenging yourself to consider somebody else's point of view and what that might be rooted in and the comments that other parties might have. … You can gain a little bit more understanding and practice trying to understand that perspective that they have.
>
> (classroom discussion)

In *Full Spectrum Resistance* McBay names "cooperation among different 'constituencies' and across traditional barriers (including ethnic lines, militants and moderates, aboveground and underground), including through alliances and coalitions" (127), as a vital component of effective organizing and activism. Collective Forum Theatre exemplifies this principle of cooperation and complementary strategizing in a fictional setting, where potential oppositions can be explored. One student reflected on the exercise and the benefits of this style of Forum Theatre, specifically in relation to strengthening social movements:

I think workshops like this, they can provide a solid form of scaffolding … not only to discuss things but actually put ourselves in these situations to try and experiment, to play off each other, and see how some of these disparate ideas might interact, and how these ideas might play out when we use them, instead of keeping it in this box, this kind of closed environment with all like-minded people. [The ideas] are not necessarily going to face the same kind of resistance as they would in the outside world. So, it's kind of a safe space to … workshop these ideas to provide some simulation about how we can best interact with the outside world, without necessarily putting ourselves in direct danger at that point.

(classroom discussion)

Another student noted that Collective Forum Theatre brings a variety of lived experiences and opinions together to debate respectfully, and to devise maximally effective strategies through collaborative means:

This [exercise] made it possible for us to bring all our own lived experiences plus the imagined lived experiences of someone else. … Instead of [someone's idea] being shot down, it was challenged and strengthened, so that every idea was as strong as it can be moving forward, so if we then did take that to the next step and we started some form of climate protest or a climate action group or something, our ideas would be as strong as we could make them. We were constantly challenging each other in a productive way.

(classroom discussion)

This reflection reminded me of an Audre Lorde quote I use often in my own theorizing, and one that I introduce to students on the first day of whatever applied theatre programme or class I facilitate. She advocates for building networks of solidarity: "[O]ur future survival is predicated upon our ability to … devise ways to use each other's difference to enrich our visions and our joint struggles" (122). As the student describes, because the exercise is theatrical, there is endless possibility for the number and type of perspectives and experiences that can be represented so as to develop a complex and informed set of strategies.

There is potential in Collective Forum Theatre to stimulate more complex thinking and future action in social movements, including a nuanced realization of the priority to engage in solidarity and to make space for the leadership of marginalized peoples. Applied theatre scholars and educators may find such a stance useful ideologically, to participate in envisioning and modelling radical futures while collaborating with progressive movements to shift the balance of power in this era of extreme socio-economic disparity and ecological turmoil. One student emphasized the need to raise up marginalized voices in Collective Forum Theatre:

We came up to this solution that left me feeling like I had a tangible strategy that I could use outside of this studio space, and had we not focused on those perspectives ... that we don't normally get a chance to hear from, ... I feel like we would have just done ... what's expected around climate change [instead of] giving power to the people that are already at the forefront of these issues. It was kind of redistributing that into these perspectives that do play a major part in this ... and what can we do to incorporate that into our everyday lives.

(classroom discussion)

Choudry notes that recognizing the variety of ways in which people participate on their own terms in social movements is both energizing and efficacious: "As people, we need to feel that we have some agency and can take action ... and that we can appreciate that different people may well have different roles and different skills to offer" (13). One student recounted how their different personal assets came into play organically over the course of the Forum exercise:

It was also interesting how we all kind of naturally fell into our personal roles that we already have. At the end, you were like, "you're usually the negotiator, you're the logistics person" ... So, it's interesting for us to see how we all have these different roles that we naturally fall into, and how we use them together collaboratively to try and create these different solutions. It was organic.

(classroom discussion)

Not only did we have a variety of lived experiences in the activity, as well as a strong foundation of social movement scholarship, we also had individuals with specific attributes that complemented the process of strategizing. For instance, the young woman playing the radical activist is a visionary; the dairy farmer is an advocate; the corporate employee is a mediator; the cashier is a tactician; the reporter stands for militant direct action; the university student expresses complex emotional and spiritual dynamics. These roles mirror the components and principles espoused by McBay in his description of what social movements need today in order to achieve success:

[W] need to see integrity in action. Where we have horizontal hostility, we need solidarity. Where we have loneliness and withdrawal, we need community engagement. Where we have symbolic actions and vague goals, we need effective action and concrete objectives. And where we have emotional suppression, we need ways of directing legitimate feelings into world-changing work.

(49)

Collective Forum Theatre establishes a site in which diverse identities, lived experiences and strategic assets can accumulate and maximize the efficacy and strength of movements for justice.

Practically speaking...

- Next steps could include Forum Theatre training and review of the texts provided below for integration in curriculum. Feel free to reach out to me for consultation, or if you would benefit from a sounding board.
- Focus on accountability and systems: Who are the people in power that (re)produce structures of oppression? Prioritize creating a list of specific multi-level demands to make part of the Forum Theatre scene.
- Conduct research on voices from communities that are most impacted prior to engaging in Forum Theatre. In particular, seek out knowledges from those that are not represented within the classroom. Emphasize the need for leadership from the most marginalized.
- Consider a central strategy with a diverse array of tactics (see McBay). Along these lines, develop plans for how people can participate in actions in a variety of ways to enhance motivation, commitment, mobilization and sustainability.

Notes

1 See for examples: Franklin, 2014; Gordon, n.d.; Kwon, 2013; Taft, 2011.
2 See for examples: Gordon, n.d.; Noguera and Canella, 2006; Taft, 2011.
3 See for examples: Bessant and Grasso, 2018; Clay 2012; Pickard and Bessant, 2018; Taft, 2011.
4 Paulo Freire was one of the foundational practitioners and scholars in critical pedagogy who based his far-reaching literacy campaign in the experiential knowledge of impoverished Brazilians beginning in the 1960s.
5 See for examples: Carruthers, 2018; Dixon, 2014; Gaztambide-Fernández, 2012; Horton and Freire, 1990; Jobin-Leeds and AgitArte, 2016; Smucker, 2017; Tools for Change, web source; Walsh, 2016.
6 I recommend McBay's (2019) *Full Spectrum Resistance: Building Movements and Fighting to Win* as a crucial resource for developing a knowledge of the history and complexity of the strategies of social movements.
7 In his foundational work from 1974, *Theatre of the Oppressed,* Augusto Boal explains the ideology and purpose behind the term "spect-actor." "In order to understand this poetics of the oppressed on must keep in mind its main objective: to change the people – 'spectators,' passive beings in the theatrical phenomenon – into subjects, into actors, transformers of the dramatic action. … [T]he poetics of the oppressed focuses on the action itself: the spectator delegates no power to the character (or actor) either to act or to think in his place; on the contrary, he himself assumes the protagonic role, changes the dramatic action, tries out solutions, discusses plans for change – in short, trains himself for real action. In this case, perhaps theater is not revolution in itself, but it is surely a rehearsal for the revolution" (Boal 122). As co-contributor Derek Davidson notes, while Boal described the ideological function of the "spect-actor" in *Theatre of the Oppressed,* he did not actually use the term until he came to write *Games for Actors and Non-actors,* first published by Routledge in 1992.

Works Cited

Abdalla, Nadine. "Youth Movements in the Egyptian Transformation: Strategies and Repertoires of Political Participation." *Mediterranean Politics,* vol. 21, no. 1, 2016, pp. 44–63.

Angus, Charlie. *Children of the Broken Treaty.* University of Regina Press, 2015.

Berardi, Franco. *Futurability: The Age of Impotence and the Rise of Possibility.* Verso, 2017.

Bessant, Judith, and Maria T. Grasso. *Governing Youth Politics in the Age of Surveillance,* Routledge, 2018.

Braxton, Eric. "Youth Leadership for Social Justice: Past and Present." In *Contemporary Youth Activism: Advancing Social Justice in the United States,* edited by Jerusha Conner and Sonia M. Rosen. Praeger, 2016, pp. 25–38.

Brucato, Ben. "An American Exception: The Counter-Insurrectionary Function of the Color Line." In *Why Don't the Poor Rise Up? Organizing the Twenty-First Century Resistance,* edited by M. Truscello and A. Nangwaya. AK Press, 2017, pp. 45–60.

Carruthers, Charlene A. *Unapologetic: A Black, Queer, and Feminist Mandate for Radical Movements.* Beacon Press, 2018.

Choudry, Aziz. *Learning Activism: The Intellectual Life of Contemporary Social Movements.* University of Toronto Press, 2015.

Clay, Andreana. *The Hip-Hop Generation Fights Back: Youth, Activism, and Post-Civil Rights Politics.* New York University Press, 2012.

Desai, Chandni. "Shooting Back in the Occupied Territories: An Anti-Colonial Participatory Politics." *Curriculum Inquiry,* vol. 45, no. 1, 2015, pp. 109–128.

Dixon, Chris. *Another Politics: Talking Across Today's Transformative Movements.* University of California Press, 2014.

Franklin, Sekou M. *After the Rebellion: Black Youth, Social Movement Activism, and the Post-Civil Rights Generation.* New York University Press, 2014.

Freire, Paulo. *Pedagogy of Solidarity.* Left Coast Press, 2016.

Gallagher, Kathleen. "Responsible Art and Unequal Societies: Towards a Theory of Drama and the Justice Agenda." In *Drama and Social Justice,* edited by Kelly Freebody and Michael Finneran. Routledge, 2016, pp. 53–66.

Gamber-Thompson, Liana, and Arely M. Zimmerman. "DREAMing Citizenship: Undocumented Youth, Coming Out, and Pathways to Participation." In *By Any Media Necessary: The New Youth Activism,* edited by Henry Jenkins, Sangita Shresthova, Liana Gamber-Thompson, Neta Kligler-Vilenchik and Arely M. Zimmerman. New York University Press, 2016, pp. 186–218.

Gaztambide-Fernández, Rubén A. "Decolonization and the Pedagogy of Solidarity." *Decolonization: Indigeneity, Education & Society,* vol. 1, no. 1, 2012, pp. 41–67.

Gomes, Daniela, and Jenaya Khan. "Our Issues, Our Struggles: A Conversation between Activists Daniela Gomes and Janaya Khan." *World Policy Journal,* 2016, pp. 47–56.

Gordon, Hava Rachel. *We Fight to Win: Inequality and the Politics of Youth Activism,* Rutgers University Press, 2010.

Harney, Stefano, and Fred Moten. *The Undercommons: Fugitive Planning & Black Study.* Minor Compositions, 2013.

Horton, Myles and Paulo Freire. "Educational Practice." In *We Make the Road by Walking: Conversations on Education and Social Change.* Temple University Press, 1990, pp. 145–173.

Jaramillo, Nathalia E. "The Art of Youth Rebellion." *Curriculum Inquiry,* vol. 45, no. 1, 2015, pp. 93–108.

Jobin-Leeds, Greg, and AgitArte. "Introduction: Inspiration from Stories." In *When We Fight We Win! Twenty-First-Century Social Movements and the Activists That Are Transforming Our World.* The New Press, 2016, pp. xiv–xxi.

Kennelly, Jacqueline. *Citizen Youth: Culture, Activism, and Agency in a Neoliberal Era.* Palgrave Macmillan, 2011.

Klein, Naomi. *NO is Not Enough: Resisting the New Shock Politics and Winning the World We Need.* Alfred A. Knopf Canada, 2017.

Kwon, Soo Ah. *Uncivil Youth: Race, Activism, and Affirmative Governmentality.* Duke University Press, 2013.

Lesko, Nancy. *Act Your Age! A Cultural Construction of Adolescence.* Routledge, 2012.

Liu, Eric. *You're More Powerful Than You Think: A Citizen's Guide to Making Change Happen.* PublicAffairs, 2017. [Social Issues Theatre & Community Engagement Course Resource].

Lorde, Audre. *Sister Outsider.* Crossing Press, 1984.

Martinez, Xiuhtexcatl. "Bioneers Conference Speech."2014. www.youtube.com/user/earthguardiankids (accessed 18 November 2016).

McBay, Aric. *Full Spectrum Resistance: Building Movements and Fighting to Win.* Seven Stories Press, 2019.

Neelands, Jonothan. "Democratic and Participatory Theatre for Social Justice: There has Never been a Famine in a Democracy. But There Will Be." In *Drama and Social Justice,* edited by Kelly Freebody and Michael Finneran. Routledge, 2016, pp. 30–39.

———. "Taming the Political: The Struggle Over Recognition in the Politics of Applied Theatre." *Research in Drama Education: The Journal of Applied Theatre and Performance,* vol. 12, no. 3, 2007, pp. 305–317.

Noguera, Pedro, and Chiara M. Canella. "Youth Agency, Resistance, and Civic Activism: The Public Commitment to Social Justice." In *Beyond Resistance! Youth Activism and Community Change: New Democratic Possibilities for Practice and Policy for America's Youth,* edited by Shaun Ginwright, Pedro Noguera, and Julie Cammarota. Routledge, 2006, pp. 333–347.

Pickard, Sarah, and Judith Bessant. *Young People Re-Generating Politics in Times of Crises.* Palgrave Macmillan, 2018.

Prendergast, Monica and Juliana Saxton. *Applied Theatre: International Case Studies and Challenges for Practice.* Intellect, 2009/2016.

Sahni, Urvashi. *Reaching for the Sky: Empowering Girls through Education.* Brookings Institution Press, 2017.

Slater, Graham B., and Briggs, C. Bradford. "Standardization and Subjection: An Autonomist Critique of Neoliberal School Reform." *Review of Education, Pedagogy, and Cultural Studies,* vol. 37, no. 5, 2015, pp. 438–459.

Smucker, Jonathan Matthew. *"The We in Politics." In Hegemony How-To: A Roadmap for Radicals, Chico, CA and* Edinburgh, Scotland, AK Press, 2017, pp. 209–232.

Taft, Jessica K. *Rebel Girls: Youth Activism and Social Change Across the Americas.* New York University Press, 2011.

Tools for Change. "1. Common Behavioral Patterns that Perpetuate Power Relations of Domination, 2. Concepts for Making Justice & 3. Getting to the Heart of the Matter." www.toolsforchange.net/resources/

Walsh, James. "Denver's Romero Theater Troupe: Welcoming Working-Class Voices in Higher Education and Revitalizing Class-Based Activism through Organic Theater." *Labor Studies Journal,* vol. 41, no. 1, 2016, pp. 114–134.

Ware, Syrus M. "Art Creates Change: The Kym Pruesse Speakers Series Featuring Black Lives Matter Toronto." Ontario College of Art and Design. 26 October 2016.

2 Strategies for climate crisis adaptation

Bringing Indigenous and Western knowledge systems together through theatre

Lara Aysal

This chapter is rooted in Denzin and Lincoln's proposal, articulated in their *Handbook of Critical and Indigenous Methodologies*, to create oppositional performance disciplines by asking to what extent bringing a variety of voices from university and community settings together around a curriculum could facilitate an opportunity to build "radical utopian spaces within public institutions" (Denzin and Lincoln 17). In this chapter I frame "radical utopian spaces" as places of coming together within public institutions, such as educational and/or community settings. I attempt to build oppositional performance pedagogies by opening a discussion about how such utopian spaces, within educational and community settings, could be built using the tools of theatre for discussions on climate change adaptation.

While other chapters in this book focus on ways to engage with climate change, this chapter focuses on processes of *adaptation* to the climate crisis. In his book *Adaptation to Climate Change*, Pelling explains that "adaptation is a social and political act; one intimately linked to contemporary, and with the possibility for re-shaping future, power relations in society" (3). I choose to define radical utopianism through Pelling's definition of adaptation, as a step towards re-shaping the future. While articulating approaches based on adaptation, it is essential to note that climate change adaptation and mitigation go hand in hand. However, during the United Nations Framework Convention on Climate Change (UNFCCC) part of the Rio Summit in 1992, the primary focus was only on mitigation (reducing greenhouse gas emissions). While mitigation was already defined in Rio, adaptation still is not clearly defined today (see the first chapter of Mark Pelling's book *Adaptation to Climate Change: From Resilience to Transformation*, 2010). According to Hasson et al. (*Climate Change in a Public Goods Game*, 2010), in the past two decades, adaptation has been at the centre of climate change debates due to the global North's failure to implement mitigation policies since 1992. In today's context, as Thunberg puts it, "our house is on fire, and change is coming whether you like it or not" (Thunberg). We do not have time to wait for climate change policies and mitigation strategies to be implemented. Today, adapting to climate change is not a choice for many people: by mid-July 2020, a quarter of Bangladesh had been flooded, affecting at least 1.5 million people.[1]

The climate crisis disproportionately puts marginalized populations at risk: it intensifies the difficulties already affecting vulnerable communities, including political and economic marginalization, loss of land and resources, human rights violations, discrimination and unemployment. Due to challenges of such magnitude, human cooperation at an unprecedented level is required to achieve both incremental and transformational climate crisis/change adaptations.[2] Even though the complexity of the climate crisis prompts a high degree of uncertainty about what the future might bring, it is important to acknowledge that this critical moment in history fosters an opportunity for people all over the world to share, practise and learn from each other about ways in which we can build a militant utopianism for taking action towards alternatives; to transform current realities into collaborative processes for positive change.

Climate change is a crisis of justice and human rights, as much as it is an environmental crisis. Extreme weather-related catastrophes are interrelated with political and economic marginalization, loss of land and resources, discrimination and unemployment. Löwy points out the exponential growth, based on capitalist logic, is what is threatening human life on this planet: "we are facing a crisis of civilization that demands radical change" (17). To take action towards radical transformative change in post-secondary theatre education, it is necessary to move beyond what we know about the climate crisis and start building vigorous pedagogical dialogues to create opportunities within public institutions for previously unheard voices. Through these radical utopian spaces, further experiments and rehearsals on how to approach transformational social change could be discussed. As many have argued, art may play an instrumental role in such discussions.[3] Such dialogue is one of many ways to close the gap between "what we know and what we actually do about climate change" (Galafassi et al. 73).

Western knowledge has been at the centre of how we approach climate change. However, partnerships between art and different knowledge systems could provide an opportunity to imagine, organize and take action in building capacity for adaptation to climate change.[4] Applied theatre, as a process of live, collaborative of storytelling, can become a medium for dialogue between Western and Indigenous pedagogies that could build collaborations for action on the crisis. In this chapter, employing the guidance of Etuaptmumk/Two-Eyed Seeing (E/TES),[5] I ask how students and communities could utilize tools of applied theatre to bring Western and Indigenous knowledge systems together as a form of militant utopianism for radical action on climate change.

In this chapter I first discuss the eco-pedagogical approach[6] to climate justice, education and praxis, specifically focusing on cognitive praxis[7] and its criticism from a non-Western perspective. Second, I bring applied theatre into the conversation as a collaborative tool that could bring together Western and Indigenous ways of knowing; that could prompt collaborations for action on climate change adaptation. And finally, I address the topic of actual curriculum design: taking into consideration the limitations of mainstream science to deal with the diversity and complexity[8] of climate change, and the need to connect knowledge with action, I explore ways to invite E/TES into the conversation to try and build a transdisciplinary framework.

Eco-pedagogy and Western epistemology

Environmental education in the era of the climate change demands radical praxis-based pedagogies. According to Khan in *Critical Pedagogy: Ecoliteracy and Planetary Crisis*, environmental education is facing a crisis because it has failed to deliver a critical perspective on systemic anthropogenic destruction, which has been supported by an economic status quo based on infinite growth (1–33). Action-based pedagogies must critique so-called sustainable education systems that encourage profit-based relations, and aim to preserve current social and political systems. The eco-pedagogy movement and the steps towards building a "cognitive praxis" offer a holistic approach that may foster radical utopian spaces.

Eco-pedagogy is both a Freirean discourse and an educational movement that sprang from the Earth Charter in 1992, at the first Earth Summit in Rio de Janeiro, Brazil. The aim of the Earth Charter was to make a systematic policy statement about educational concerns through ethical and ecological lenses (Khan 12). Eco-pedagogy suggests systems change by making economic, social and cultural structural changes.

Combining Freirean aims of the "humanization" (18) of experience, and the achievement of a just and free world, eco-pedagogy "militantly" opposes neoliberalism and "attempts to foment collective ecoliteracy and realize culturally relevant forms of knowledge grounded in normative concepts such as sustainability, planetarity, and biophilia, on the other" (ibid.). Eco-pedagogy aims to critically transform education through eco-socialist discourse. It invokes Herbert Marcuse's ecological politics on education, focusing on how to overcome the climate crisis through the "creation of revolutionary struggle and the search of new life sensibilities capable of transcending the nature/culture dichotomy" (22). Even though Marcuse died in 1979, years before climate change had emerged as a crisis, his educational theory

> was essentially linked to the ecological problem of human and non-human relations due to his understanding that education is a cultural activity, and that in Western history such culture has systematically defined itself against nature in both a hierarchically dominating and repressive manner.
>
> (137)

Marcuse's ideas can be enlisted to move eco-pedagogy away from the Freirean dichotomy of the human/non-human. For his part, Marcuse considers education as a radical act of revolution that is embedded in political consciousness and collective struggles (ibid.). Khan thus outlines eco-pedagogy as a "[s]ociopolitical movement that acts pedagogically throughout all of its varied oppositional political and cultural activities" (23). According to Eyerman and Jamison, eco-pedagogy is therefore an environmental movement that has the capacity to become a socially constructed force that engages with society pedagogically, negotiates with dominant institutional powers and forms a collective cognitive praxis. In defining cognitive praxis in his article *Social Movements and Science: Cultural Appropriations of Cognitive Praxis,* Jamison points out that he and Ron Eyerman combined three subject areas of environmental movements, world-view assumptions (cosmology), criteria for technical change (technology) and organizational forms (organization), "into an integrative cognitive praxis to provide an important part of what other movement theorists have termed 'collective identity'" (47). Jamison and Eyerman define collective identity in these social movements "as spaces to experiment and construct Western scientific knowledge" (47). These public spaces formed the basis for "innovative forms of cognitive praxis combining new worldview or cosmological assumptions with alternative organizational forms and technological criteria" (ibid).

As Jamison points out in *The Making of Green Knowledge: Environmental Politics and Cultural Transformation* (2001), the collective identity of social movements has historically transformed into the "professionalization and specialization" of environmental industrialization and the development of sustainable energy (16–44). It can be understood that the environmental movement of the 1970s and 1980s has shaped the environmental sciences and transformed the institutions that we engage with today. The cultural translation of this process has not built a dialogical relation between the social movements and the development of sustainable energy. As Jamison affirms in *The Making of Green Knowledge: Environmental Politics and Cultural Transformation,* the commercialization of environmental movements has led to a profit-based approach. Cognitive praxis argues that hybrid identities can exceed disciplinary boundaries between academics and activists and build a dialogical relation for cultural transformation of knowledge through collective identities. Eco-pedagogy is a movement that prioritizes cognitive praxis by bringing education to the centre of political action, as a "revolutionary critical pedagogy based in hope that can bridge the politics of the academy with forms of grassroots political organizing capable of achieving social and ecological transformation" (Martin qtd. in Khan 349).

The tools of applied theatre could be used to bring together the notion of cognitive praxis with the approaches of eco-pedagogy in order to inspire radical action on climate change adaptation in the classroom. In order to move away from the commercialization of environmental movements, it is necessary to use a holistic approach to environmental education, as Khan argues: "[E]ducation today must involve the mind and the body, reason and imagination, intellectual and the instinctual needs, because our entire existence has become the subject/object of politics, of social engineering" (85). However, as Bowers explains in *Re-Thinking Freire,* during the conference entitled "Freire and Beyond" (Smith College and the University of Massachusetts in 2000), activists from countries in the developing world re-investigated "Freire's vision of empowerment, which they initially interpreted as a non-colonizing pedagogy" (2). However, during the conference they had a chance to engage with Indigenous culture, and this is how "they became aware that Freire's ideas are based on Western assumptions and that the Freirean approach to empowerment was really a disguised form of colonization" (2).

Taking Bowers' point of view into consideration, it is possible to argue that performance – and arts-based action share common ground and multiple tools with Indigenous ways of knowing.[9] I will briefly remind readers what applied theatre is, and then suggest how a common language with Indigenous epistemologies could be built for environmental education.

Applied theatre

Applied theatre is a performance-based inquiry that takes place with, by and for communities as a form of artistic practice. Applied theatre refers to theatre not usually made within traditional theatre buildings, but rather with and within communities.[10] It requires the participants to work together to find aesthetic solutions to problems emerging in the creative and social processes, which enables them to use their imagination while interacting with one another. Applied theatre is an umbrella term often used to describe many different practices, including Augusto Boal's Theatre of the Oppressed (considered by other writers in this volume), classroom drama, theatre-in-education, community-based performance, prison theatre, Theatre for Development, political theatre, social theatre, educational theatre, engaged performance and Theatre and Social Change.[11] It is different from Western theatre production in its emphasis on community-engaged participatory processes, and often focuses on accomplishing a particular goal in community settings. These goals emerge from lived experience, individual or collective values and traditions, and focus on contested issues within or around communities. Applied theatre accommodates possibilities for dialogue and praxis through participation-based creative approaches,[12] and is able to access that which is otherwise out of the reach of science by giving primacy to experience-based collective consciousness.

In his lecture *On Sensitivity Arts, Science and Politics in the New Climatic Regime*, Latour points out that that theatre is an ideal milieu to

> replay alternative spatio-temporal frames, to make non-speaking entities speak, to explore alternative plots, to assemble the public (audience) in a different way, to imitate the ways models (climate models) are built-in science by adding new variables and explore alternative outcomes, to explore ways for the audience to change their attachments to the issues.

> (Latour)

Adapting Latour's exploration of climate change within the realm of theatre, instrumental community-based principles of applied theatre, such as collaboration, participation and action, through storytelling and role-playing, can contribute to this dialogue, and further engage with transformative social change.

Averill argues that the human-rights perspective on climate change can help people to focus on culture and tradition. Through these stories, the discussion may move beyond the scientific debates and focus on values and traditions that are equally affected by climate change. From a values-based point of view, scholars have argued that "adaptation is recognized as an effort to reduce climate-related risks and to protect things we value or to keep risks to valued objectives at a tolerable level" (Tschakert 2). Thus, the concept of value within the context of climate change adaptation is identified as a broad term including, but not limited to, culture and ways of knowing. Increasingly we observe a discursive shift towards an understanding that both the impacts of and solutions to climate change are deeply mediated by culture.[13] Indeed, as all the contributors to this volume believe and argue, the arts can play a crucial role in understanding modes of thinking about climate change and encouraging people to imagine a world beyond the given present.[14] Specifically, applied theatre can become an effective tool to communicate about, explore and build the ground for taking active steps towards climate change adaptation. As a form of storytelling, it can function as a creative method of approaching the processes of climate change adaptation from the perspective of cultural transformation, building dialogue between various knowledge systems. Taking action in this crisis requires a transdisciplinary process aiming at sharing experiences for co-creating knowledge.[15]

Indigenous ways of knowing

As is increasingly known across a range of institutions and discourses, Indigenous knowledge is rooted in observations of and interactions with local ecosystems accumulated over generations. Indigenous communities have been interpreting and reacting to the impacts of climate change, drawing on traditional knowledge to find solutions and initiate dialogue, which may help society at large to cope with changes. However, the stewardship role of Indigenous peoples on issues concerning climate change is being underestimated by the Western world.[16] Even though eco-pedagogy and cognitive praxis point to possibilities regarding how applied theatre could facilitate dialogue within the Western cultural practices, Bowers' perspective outlines the uncertainty of critical pedagogy to recognize the profound difference in cultural ways of knowing, including those in orally based cultures (2, 8). Western knowledge has been at the centre of the conversation on adaptation processes, leaving cultural aspects out of the conversation. Within this critical perspective, applied theatre could endorse multi-cultural, knowledge-based prospects that can both move forward with, but also move beyond, the Western knowledge framework.

According to the *Science First Peoples Teacher Resource*, a production of the First Nations Education Steering Committee, "Indigenous Science is the knowledge of Indigenous peoples, including scientific and evidence-based knowledge, which has been built up over thousands of years of interaction with the environment" (Bernabei et al. 7). It is also defined as a "holistic and relational knowledge rooted in place and contained in language" (ibid.). From a cultural and cosmological perspective, it points to the interconnectedness of all living things, and "this principle governs all relationships. It governs relationships between all human beings, between all other forms of life, and binds all together within one continuous web of creation" (Stewart-Harawira 155). And the relationship of living and non-living things is also interconnected at every level and forms part of the whole that signifies existence. Within the teachings of knowledge on the continuous web of creation, and how to exist within it, Indigenous knowledge borrows elements familiar to arts-based practices: storytelling, dance and singing have been integral to transferring knowledge for thousands of years. Kovach notes, in *Indigenous methodologies: Characteristics, conversations, and contexts*, that

> stories hold within them knowledges while simultaneously signifying relationships. They are active agents within a relational world, pivotal in gaining insight into a phenomenon. Oral stories are born of connections within the world, and are thus recounted relationally. They tie us with our past and provide a basis for continuity with future generations.
>
> (94)

The interconnectedness of Indigenous ways of knowing with storytelling draws a pedagogical analogy with theatre, as both of them entail embodied observation and practice. Storytelling has the flexibility to move through time and space, and transfer, transform and analyse knowledge that might be beyond the reach of Western science. The E/TES approach proposes a holistic process-based co-creation of knowledge that equally acknowledges Western and Indigenous systems.

The E/TES approach

According to Bartlett et al., E/TES "is a gift of multiple perspectives" (335). It is a learning process that invites us

> to see from one eye with the strengths of Indigenous knowledges and ways of knowing, and from the other eye with the strengths of Western knowledges and ways of knowing, and to using both these eyes together, for the benefit of all.
>
> (ibid)

The E/TES approach, defined by Elder Albert Marshall (Mi'kmaw Nation), could provide the guiding principle for a process that stems from the

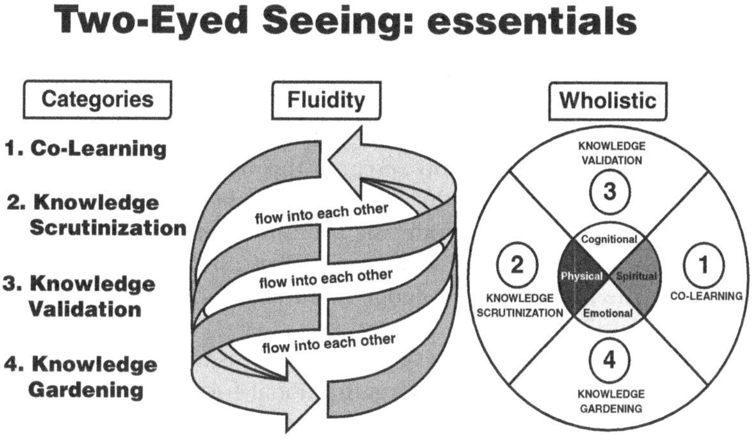

Two-Eyed Seeing: essentials

Figure 2.1 Essentials of Etuaptmumk/Two-Eyed Seeing for Knowledge Gardening.[17]

collaboration between knowledge systems and applied theatre. In what follows the main principles of the E/TES approach are taken into account as a guide. These steps are defined by Bartlett and Elder Marshall as co-learning, knowledge scrutinization, knowledge validation and knowledge gardening.

With the guidance of the E/TES approach, I propose to utilize tools for applied theatre in the classroom environment to bring these knowledge systems together in order to rethink how to approach the climate crisis from a dialogical and action-based perspective. This process would require the involvement of knowledge holders (from Indigenous and Western knowledge systems) to contribute to the co-teaching process. In short, transdisciplinary framework, bringing Western knowledge and Indigenous knowledge together with applied theatre, can amplify how we understand what we value, while co-producing knowledge through a dialogical pedagogy.

Bringing the pieces together: transforming and weaving cognitive praxis

Transdisciplinary research focuses on producing knowledge to understand complex problems and their relation to societal transformation.[18] In his six-volume *La Méthode*, Edgar Morin introduces transdisciplinary research as a unity accompanied by diversity. The complexity of the relationship between our species and the ecosystem cannot be grasped from a problem/solution-based dichotomy. As Morin points out, "the whole is more and less than the sum of its parts" (112). These systems, including environmental change and movements for climate justice, are intertwined in a complex structure that needs multiple voices to initiate action. Aligned with this research, the E/TES approach embodies the complexity of climate change and the multiplicity of perspectives that should be taken into account. I adopt Nicolescu's definition of transdisciplinarity: "at once between the disciplines, across the different disciplines, and beyond all discipline, aiming at the understanding of the present world, of which one of the imperatives is the unity of knowledge" (44). Lang and colleagues conceptualize a transdisciplinary process for sustainability research through three phases, defined as "[c]ollaboratively framing the problem and building a collaborative research team (Phase A); co-producing solution-oriented and transferable knowledge through collaborative research (Phase B); and (re-)integrating and applying the produced knowledge in both scientific and societal practice (Phase C)" (27).

As discussed above, the aim of cognitive praxis is to combine cosmological assumptions with alternative organizational forms and technological criteria. The aim of this combination is to generate a societal pedagogy that negotiates with dominant institutional powers. I attempt to build a transdisciplinary curriculum guideline that could enrich the co-production of knowledge through the guiding principles of Lang and colleagues on transdisciplinary research processes, and the E/TES approach. They are combined as follows:

1. and 4. Co-learning and Knowledge Gardening through Addressing Curiosity: framing the problem through workshop/seminar-lectures with knowledge holders on climate change.
2. Knowledge Scrutinization through Taking Action: re-integrating and applying the produced knowledge in both scientific and societal practice.
3. Knowledge Validation through Organizing Curiosity: co-producing solution-oriented and transferable knowledge through art-based research methods for community engagement. Focusing on producing knowledge to understand complex problems and their relation to societal transformation.[19]

Curriculum structure

A transdisciplinary collaboration among scientific environmental knowledge, traditional ecological knowledge and applied theatre could facilitate possibilities for a dialogical encounter on climate change processes through two steps: (1) transforming and weaving materials from scientific environmental knowledge and traditional environmental knowledge into theatrical performances in a classroom setting; and (2) presenting the outcomes, that is, performances, to communities to spark a participatory dialogue between students and communities. Co-creation of knowledge through collaborations could evoke critical thinking and promote social learning approaches as a form of militant utopianism.

An interdisciplinary course could be developed to bring together students from various departments focusing on climate change, including political ecology, ethnoecology, ecological restoration, Indigenous episte-mologies and applied theatre, to explore Western and Indigenous points of view on climate change. Lectures written, organized and analysed by students could guide the transformation of knowledge into perfor-mance-based inquiries. Applied theatre can evoke a sense of agency in students, who, engaging critically with the knowledge gained from the lec-tures, could develop scenarios and short plays on issues arising from them to be performed/workshopped in community settings.

The course could be divided into two sections. In the first section, knowl-edge holders (as an advisory committee) from Western and Indigenous knowledge systems could give lectures on various perspectives on climate change and adaptation to it. During the second section, students could realize the knowledge they gathered through dramatic approaches. They could utilize improvisation to articulate what they had learned (data col-lection); how they frame knowledge through tools of theatre (analysis); and how they will present it to others through performances[20] (dissemi-nation). The advisory committee could guide this process by taking the role of the audience to challenge students to bring multiple perspectives into the conversation. Students would collect feedback and revise their per-formance scores in accordance with the guidance of the advisory commit-tee. Embodying knowledge as performance could take place, aiming to give agency to students to critically engage with how—and whether—bringing two knowledge systems together, with the guidance of the E/TES approach, is possible. Through this structure, knowledge holders, students and com-munity members could engage with the knowledge that is guided and trans-formed with/by/for all stakeholders. This collaboration could contribute to co-creating knowledge for environmental education by connecting multiple disciplines with imaginative ideas and communities.[21]

Concluding by returning to the beginning

It is May 1, 2020 as I write this. The world is in a totally different place from where it was when I began to write this chapter. Today, in the midst of a pandemic, we are going through unprecedented times and trying to understand what is happening to the world, as we try to find ways in which we can connect without contact. We are living under lockdowns, a dystopic reality for the privileged, and an everyday fact for many people around the world who do not have access to the same privileges. Perhaps now is the right time to ask how applied theatre can bring Indigenous and Western knowledge systems together as a form of militant action on the climate crisis. Just as the pandemic poses risks to our very existence today, it also provides a key to how the climate crisis is interrelated to a broader complexity of the future to come. These uncertain days also bring oppor-tunities to acknowledge that this critical moment in history fosters possi-bilities to share, practise and learn from each other about ways in which we can take action towards alternatives, transforming reality into collaborative processes.

While Western science is racing against time to find a vaccine to fight the virus, Indigenous knowledge keeper/language consultant Jeff Wastisicoot, from Crosslake, Manitoba, on Treaty 5 territory, warns us that breaking certain laws, such as eating animals that ought to be avoided, disconnecting from the land and its medicines, leads to pathogenic diseases.[22] Wastisicoot points out that the preparation practices to avoid these diseases are embedded in Indigenous knowledge and exist within the language of the land. We need to create balance through both Western and Indigenous knowledge systems. Also, Isaac Murdoch, from Serpent River First Nation, advises that certain practices, such as "dish with one spoon" (a teaching for trade and community support), and spirit stories teach us how to understand and react to such diseases.[23] If elders and knowledge keepers are pointing out that laws have been broken, and the medicines that Western and Indigenous knowledges utilize to cure diseases are being compromised through deforestation, pollution, extraction and so on, what actions are we going to take to build collaborative practices in response to such crises? In between global pandemics, local disasters and economic and social devastations, it is apparent, more than ever, that we need to find tools and inspirations that could bring about the future we want to live in.

Practically speaking...

E/TES-informed applied theatre processes

1   **and 4. Co-Learning Collaboration**
    Artists/students collaborate with elders and scientists to discuss what we value in relation to climate change adaptation.

2   **Knowledge Scrutinization—Applied Theatre/ Devising/Rehearsing**
    Analysing knowledge from elders and scientists by forming stories with artists/students. Devising stories with applied theatre practitioners.

3   **Knowledge Validation—Action/Acting/Forum**
    Sharing the devised stories with elders and scientists.

4   **and 1. Knowledge Gardening / Co-Learning- Reflection/Action**
    Transforming the stories with feedback. Redeveloping the stories. After sharing the performances with elders and scientists, opening the work up to community members through a range of forums, including participatory processes and presentations.

Through this structure the aim will be to address the human dimension of the climate crisis in classrooms and community settings. Applied theatre would enhance engagement among all participants (students, co-researchers, advisory committee and community members). The focus would be on social learning through theatre that emphasizes in-depth analysis, ethnography, observations for critical and creative thinking and praxis-based dialogue between stakeholders.

Notes

1 It is crucial to engage in climate justice perspectives while bringing adaptation to the fore as social and political action. It is essential to identify the difference between the perspectives of the global South vs. the North. The global North refers to wealthy countries that are the primary consumers, producers and polluters of the world, formerly known as the "First World," whose economies are dependent on fossil fuel, mining and other forms of extraction. For example, 75% of the world's mining companies are based in Canada. The global South is defined as those counties that consume, produce and pollute the least. However, these countries are often hit the hardest by climate disasters. Cassegard and Thörn argue in *Toward a postapocalyptic environmentalism?* that "the whole idea of a future disaster was perceived as North-centric, considering that climate-related disasters were already occurring in many poor countries" (568). Thus, the idea that climate change might occur in the future is a North-centric perspective that disregards the fact that global South is already experiencing the devastations of climate change. Therefore, adapting to climate change is an ongoing real-time response to loss and damage, specifically for the global South. The injustice between global North and South points to the necessity for systemic change, following a radical alternative instead of a reformist pathway. I here adopt the definitions "reformist" and "radical alternatives" from Temper et al. Reformist initiatives (civil society organizations, government agencies and businesses) deal "only with the symptoms of the problem" (752), while radical alternatives "[confront] the basic structural reasons for unsustainability, inequity and injustice, such as capitalism, patriarchy, state centrism, or other inequities in power resulting from caste, ethnic, racial, and other social characteristics" (ibid).

2 See the report: *IPCC special report on the impacts of global warming of 1.5 °C. Global Warming of*, 15.

3 See Bagley, C. Educational ethnography as performance art: Towards a sensuous feeling and knowing. *Qualitative Research*, 8(1), 2008, pp. 53–72.; Barone, T., and Eisner, E. W. *Arts based research.* Sage, 2011.; Denzin, N. K. Performance ethnography: Critical pedagogy and the politics of culture. Sage, 2003.; And Fox, M. Embodied methodologies, participation, and the art of research. *Social and Personality Psychology Compass*, 9(7), 2015, pp. 321–332.

4 Further research: Finley, S. Arts-based research. *Handbook of the arts in qualitative research: Perspectives, methodologies, examples, and issues*, 2008, pp. 71–81; Cole, A. L., and Knowles, J. G. Arts-informed research. *Handbook of the arts in qualitative research*, 200, pp. 55–70; Sullivan, G. (Ed.). *Art practice as research: Inquiry in visual arts.* Sage, 2010.; Leavy, P. (Ed.). *The Oxford handbook of qualitative research.* Oxford University Press, 2014.

5 For further information: Bartlett, C., Marshall, M., and Marshall, A. Two-eyed seeing and other lessons learned within a co-learning journey of bringing together Indigenous and mainstream knowledges and ways of knowing. *Journal of Environmental Studies and Sciences*, 2(4), 2012, pp. 331–340.

6 See Kahn, R., and Kahn, R. V. *Critical pedagogy, Ecoliteracy, and Planetary Crisis: The Eco-pedagogy Movement* (Vol. 359). Peter Lang, 2010.

7 See Jamison, A. Social movements and science: Cultural appropriations of cognitive praxis. *Science as Culture*, 15(01), 2006, pp. 45–59.

8 See Pohl, C. From science to policy through transdisciplinary research. *Environmental science and policy*, 11(1), 2008, pp. 46–53.; Pohl, Christian,

Stephan Rist, Anne Zimmermann, Patricia Fry, Ghana S. Gurung, Flurina Schneider, Chinwe Ifejika Speranza et al. Researchers' roles in knowledge co-production: experience from sustainability research in Kenya, Switzerland, Bolivia and Nepal. *Science and Public Policy*, *37*(4), 2010, pp. 267–281.; Hadorn, G. H., Bradley, D., Pohl, C., Rist, S., and Wiesmann, U. Implications of transdisciplinarity for sustainability research. *Ecological Economics*, *60*(1), 2006, pp. 119–128.

9   See Denzin, N. K., Lincoln, Y. S., and Smith, L. T. (Eds.). *Handbook of critical and indigenous methodologies*. Sage, 2008.

10 See Nicholson, H. *Applied drama: The gift of theatre*. Macmillan International Higher Education, 2014.; Thompson, J. *Applied theatre: Bewilderment and beyond* (Vol. 5). Peter Lang Publishing. 2003.

11 See Snyder-Young, D. *Theatre of good intentions: Challenges and hopes for theatre and social change*. Springer, 2013.

12 See Prentki, T., and Preston, S. (Eds.). *The applied theatre reader*. Routledge, 2013.

13 See, for example, the work of Tàbara, J. D., Clair, A. L. S., and Hermansen, E. A. Transforming communication and knowledge production processes to address high-end climate change. *Environmental Science and Policy*, *70*, 2017, pp. 31–37.

14 See Yusoff, K., and Gabrys, J. Climate change and the imagination. *Wiley Interdisciplinary Reviews: Climate Change*, *2*(4), 2011, pp. 516–534.

15 See Galafassi, D. *The Transformative Imagination: Re-imagining the world towards sustainability* (Doctoral dissertation, Stockholm Resilience Centre, Stockholm University). 2018.

16 See Garcia-Alix, L. The United Nations permanent forum on Indigenous issues discusses climate change. *Indigenous Affairs*, 2008, pp. 1–23.

17 See Bartlett and Marshall, *Key essentials Etuaptmumk/Two-Eyed Seeing for Knowledge Gardening*. 2018, p. 2.

18 See Vilsmaier, U., and Lang, D. J. Transdisziplinäre Forschung. In *Nachhaltigkeitswissenschaften* (pp. 87–113). Springer Spektrum, 2014.; Lang, D. J., Wiek, A., Bergmann, M., Stauffacher, M., Martens, P., Moll, P., ... and Thomas, C. J. Transdisciplinary research in sustainability science: practice, principles, and challenges. *Sustainability Science*, *7*(1), 2012, pp. 25–43.

19 See Vilsmaier, U., and Lang, D. J. Transdisziplinäre Forschung. In *Nachhaltigkeitswissenschaften* (pp. 87–113). Springer Spektrum, 2014.; Lang, D. J., Wiek, A., Bergmann, M., Stauffacher, M., Martens, P., Moll, P., ... and Thomas, C. J. Transdisciplinary research in sustainability science: practice, principles, and challenges. *Sustainability Science*, *7*(1), 2012, pp. 25–43.

20 See Norris, J. Drama as research: Realizing the potential of drama in education as a research methodology. *Youth Theatre Journal*, *30*(2), 2016, pp. 122–135.

21 See Hulme, M. *Why we disagree about climate change: Understanding controversy, inaction and opportunity*. Cambridge University Press. 2009.

22 See Indigenous Climate Action Network. 2020, March 19, *Webinar on COVID19 and Indigenous Communities*. [Video]. YouTube. https://www.youtube.com/watch?v=K57p0gApbz4andfeature=youtu.be.

23 See the video source above: See Indigenous Climate Action Network. 2020, March 19, *Webinar on COVID19 and Indigenous Communities*. [Video]. YouTube. https://www.youtube.com/watch?v=K57p0gApbz4andfeature=youtu.be.

Works Cited

Averill, Marilyn. "Linking Climate Litigation and Human Rights." *Review of European Community and International Environmental Law*, vol. 18, no. 2, 2009, pp. 139–147.

Bartlett, Chery. "Two-Eyed Seeing an Overview of the Guiding Principle Plus Some Integrative Science." *Institute for Integrative Science & Health.* 23 August 2017. http://integrativescience.ca. With authors permission.

Bartlett, Cheryl, Murdena Marshall, and Albert Marshall. "Two-Eyed Seeing and Other Lessons Learned Within a Co-Learning Journey of Bringing Together Indigenous and Mainstream Knowledges and Ways of Knowing." *Journal of Environmental Studies and Sciences*, vol. 2, no. 4, 2012, pp. 331–340.

Bernabei, Mati, Tannis Calder, and Stephanie Sedwick. "First Nations Education Steering Committee and First Nations Schools Association." *Secondary Science First Peoples Teacher Resource Guide*, 2019.

Bowers, Chet A. *Re-Thinking Freire: Globalization and the Environmental Crisis.* Edited by C. A. Bowers and F. Apffel-Marglin. Routledge, 2004.

Burns, Danny, Blane Harvey, and Alfredo O. Aragón. "Introduction: Action Research for Development and Social Change." *IDS Bulletin*, vol. 43, no. 3, 2012, pp. 1–7.

Cassegård, Carl, and Håkan Thörn. "Toward a Postapocalyptic Environmentalism? Responses to Loss and Visions of the Future in Climate Activism." *Environment and Planning E: Nature and Space*, vol. 1, no. 4, 2018, pp. 561–578.

Denzin, Norman K., Yvonna S. Lincoln, and Linda T. Smith. *Handbook of Critical and Indigenous Methodologies.* Los Angeles, Sage, 2008.

Eyerman, Ron, and Jamison, Andrew. *Social Movements: A Cognitive Approach.* Penn State Press, 1991.

Galafassi, Diego, Sacha Kagan, Manjana Milkoreit, Maria Heras, Chantal Bilodeau, Sadhbh Bourke, and Joan D. Tàbara. "'Raising the Temperature': The Arts on a Warming Planet." *Current Opinion in Environmental Sustainability*, vol. 31, 2018, pp. 71–79.

Jamison, Andrew. *Participation and Agency: Hybrid Identities in the European Quest for Sustainable Development.* Edited by R. Paehlke and D. Torgerson. Broadview Press. 2005.

———. *The Making of Green Knowledge: Environmental Politics and Cultural Transformation.* Cambridge University Press, 2001.

Kahn, R., and R. V. Kahn. *Critical Pedagogy, Ecoliteracy, and Planetary Crisis: The Eco-Pedagogy Movement* (Vol. 359). Peter Lang, 2010.

Kovach, Margaret. *Indigenous Methodologies: Characteristics, Conversations, and Contexts.* University of Toronto Press, 2009.

Latour, Bruno. *On Sensitivity Arts, Science and Politics in the New Climatic Regime.* [Video]. University of Melbourne. 7 July 2016. YouTube. https://www.youtube.com/watch?v=hTzhTlrNBfw. [48,01]. (accessed 1 February 2020).

Levac, Leah, Lisa McMurtry, Deborah Stienstra, Gail Baikie, Cindy Hanson, and Devi Mucina. "Learning Across Indigenous and Western Knowledge Systems and Intersectionality: Reconciling Social Science Research Approaches." *Unpublished SSHRC Knowledge Synthesis Report).* University of Guelph, 2018.

Löwy, Michael. "What is Ecosocialism?" *Capitalism Nature Socialism*, vol. 16, no. 2, 2005, pp. 15–24.

Morin, Edgar. *Tome La Méthode. 1: La nature de la Nature.* Paris, Ed. du Seuil, 1977.

Nicolescu, Basarab. *Manifesto of Transdisciplinarity.* Suny Press, 2002.

Nicolescu, Basarab, Edgar Morin, and L. de Freitas. "The Charter of Transdisciplinarity." *Manifesto of trandisciplinarity.* 1994.

Pelling, Mark. *Adaptation to Climate Change: From Resilience to Transformation.* Routledge, 2010.

Stewart-Harawira, Makere. "Cultural Studies, Indigenous Knowledge and Pedagogies of Hope." *Policy Futures in Education*, vol. 3, no. 2, 2005, pp. 153–163.

Taylor, Philip. *Applied Theatre: Creating Transformative Encounters in the Community*. Portsmouth, N.H, 2003.

Temper, Leah, Mariana Walter, Iokiñe Rodriguez, Ashish Kothari, and Ethemcan Turhan. "A Perspective on Radical Transformations to Sustainability: Resistances, Movements and Alternatives." *Sustainability Science*, vol. 13, no. 3, 2018, pp. 747–764.

Thunberg, Greta. "Our House Is on Fire!" World Economic Forum, 25 January 2019. YouTube, 12:54. (accessed 1 September 2020).

Tschakert, Petra, Jon Barnett, Neville Ellis, Carmen Lawrence, Nancy Tuana, Mark New, Carmen Elrick-Barr, Ram Pandit, and David Pannell. "Climate Change and Loss, as if People Mattered: Values, Places, and Experiences." *Wiley Interdisciplinary Reviews. Climate Change*, vol. 8, no. 5, 2017, pp. 1–19.

3 Voices we carry within us

A trialogue about climate change, Indigenous ways of knowing and activism

Lara Aysal and Dennis D. Gupa in Conversation with Kirsten Sadeghi-Yekta

KIRSTEN SADEGHI-YEKTA In this chapter we investigate the processes and meanings of climate change through our artistic and academic encounters with Indigenous methodologies. This chapter is a trialogue of conversational encounters between Lara Aysal, Dennis Gupa and Kirsten Sadeghi-Yekta, as interlocutors reflecting on themes of environmental destruction, environmental justice and the overwhelming impact that climate change presents in our field sites, research and artistic works. (Sadeghi-Yekta has facilitated the conversation between Gupa and Aysal, and makes editorial reflections throughout the text.) We situate these encounters with climate change by citing references from our artistic and academic projects. Central to this conversation is how Indigenous epistemology and ontology play out in traditional performance forms and political actions. Finally, we attempt to explore the significance of Indigenous research methodologies within the context of climate change through an arts-based approach.

To start the conversation: Dennis, could you summarize the inspirations or the personal origins of your recent research questions?

DENNIS GUPA The impetus of this dissertation on climate change and applied theatre connects with my first arrival here in Canada as an MFA Directing student at the University of British Columbia (UBC). I came here to study theatre, intending to fully immerse myself in the craft of directing for the stage. I left the Philippines in September. It was sunny when I arrived in Vancouver. My host family brought me directly to Kitsilano Beach. I remember how quiet the water was. After two months of my arrival there was a typhoon that struck the Philippines. One of the theatre faculty members at the Theatre Department and Film at UBC kindly checked if my family was okay. I was clueless about what was happening. We are used to massive typhoons, and therefore I told my professor not to worry about it. My quip was not successful once my professor showed me pictures of the devastation. It was Super Typhoon Yolanda. The news agencies would describe it as "monster typhoon," or the "beast that caused the worst disaster." And I was shocked with the images of the detritus. I was cut off from the suffering of my people. It was hard. Meanwhile, the suffering on site in Leyte and Samar Provinces, the ground zeros of Super Typhoon Yolanda, was enormous. After my MFA I applied to pursue a PhD in Applied Theatre at the University of Victoria's Department of Theatre, with climate change and Indigenous ecological knowledge as a topic of enquiry, by looking at ritual performances in island communities.

KIRSTEN That was an introduction to your research. One pivotal question we always ask as researchers in arts-based approaches is, "Where are you at right now?" In other words, where has your research taken you so far and what are the future steps in your research? Could you describe where you are situated in your research at the moment? It would be good for the reader to know that you have recently finalized your fieldwork.

DENNIS That is correct. After my fieldwork, I organized the "data"[1] that my collaborators and I co-produced. These data are in the forms of stories which the elders and community members shared with me, autoethnographic sketches that I wrote during my stay on the island and archival materials that I obtained from the libraries in the Philippines. I am intrigued with the colonial historiography embedded in these archival materials. I studied these data with the goal of transforming them as performance texts, which is intellectually challenging but also an empowering creative process. I engaged with these materials as performance pieces using post-colonial perspectives to interrogate climate justice in the Philippines. I am grateful that the community gave me stories which are private, funny, sad and, yes, complex. I am currently writing my dissertation, and the biggest challenge now is how I can represent them in a written form because their nature is performance-based—they are spoken or orally shared to me. Now that I am telling you about them, I think it is necessary to share how I envision them to be: these disastrous events that the community members experience are part of the cultivation of creativity and renewal of their imagination in reconstructing their futures.

KIRSTEN Before we discuss these fascinating observations, I would first like to hear Lara's research introduction.

LARA AYSAL I am an actor. I was doing mainstream theatre in Turkey, and at one point I felt that acting was not responding to what I was searching for. I was looking for a more intimate dialogue with the audience and building a more profound and meaningful communication with people through theatre. So, I started to search for artists, communities and theatres that positioned themselves in a similar way. This led me to my involvement in guerrilla theatre and performative encounters on the streets, which turned out to be experimental encounters with the police even more so than with the people (*laughs*). Through these experiences, I started to question how creative and critical thinking could be introduced to people who are not particularly interested in the doing of theatre. Applied theatre—I would not have called it applied theatre back then—could build that dialogue. My inspiration stems from the urgency of the climate crisis. A couple of years ago my focus in my work was on human rights and injustices in various forms. I was collaborating with communities who were deeply affected by the 2011 earthquake in Van, in the Kurdish territories. Specifically, I was working with women's groups, refugees and collaborating with institutions such as youth detention centres, working on such issues as child sexual abuse. So, I was trying to collaborate with a variety of communities.

And what I came to understand, during these experiences and after doing some research, was that all of these issues of injustice are interrelated with the climate crisis; not climate change itself, as "changing climates" from an environmental perspective, but more as struggles and issues that are accumulating around climate justice. In my mind it all came to be a holistic picture, where it is all interrelated through the climate crisis. So, I found an entry point: "How do I situate this within climate justice discussions?" I found the climate crisis as a common language or a ground that helped me understand the cause and effect of neoliberalism in relation to these issues. Moreover, I started to find inspiration from movements, social justice struggles and resistances that have been going on in the Middle East, in Europe and in North America for the past ten to fifteen years. I was actively involved in some of these movements, and I was transformed forever, into who I am today. What is inspiring to me is how people are trying to push back against totalitarian regimes and create space, tactics and practices for political, social and economic collectivism and plurality. I also find inspiration from land defenders who are putting their lives on the line to protect life for all. Indigenous nations are protecting us as we speak throughout Turtle Island; village women on the Black Sea are blocking roads to protect rivers in Anatolia; and activist groups across Europe, like Ende Gelände, are occupying coal mines asking for system change, not climate change. I am becoming more and more interested in understanding what arts' involvement is or could be, in this collective language that we are trying to build, as we speak.

In my chapter[2] I explain this in detail; however, to briefly mention here: I recently started to learn about Indigenous peoples, their struggles and their histories. I also started to see how these movements feed, inform and contribute to climate justice movements all around the world. I am trying to learn from Indigenous warriors and knowledge keepers how Indigenous ways of knowing and learning are connected with the land. I am also interested in the processes of how a collective knowledge, an allyship is developing around the climate crisis through Indigenous knowledge and Western knowledge.

KIRSTEN To ask the same pivotal question to you Lara, as a researcher, at what stage in your research are you *at* the moment?

LARA My journey has been an interesting process. Currently I am transferring my PhD in Applied Theatre at UVic to Interdisciplinary Studies at UBC. I am also developing projects to bring Indigenous and Western knowledge systems together through story telling with Greenpeace Canada, Indigenous knowledge holders, artists, communities and scientists in Vancouver, and I am certain these collaborations will contribute to my research.

KIRSTEN Could you both tell me your research questions?

DENNIS "In what ways do local elders provide a perspective pertinent to ongoing debates on climate change, adaptation and mitigation?" The local elders are my interlocutors in my dissertation. I grew up experiencing typhoons and learned that they co-exist with people. They were part of my life. When Super Typhoon Yolanda happened on November 8, 2013, there was indignation towards the typhoon. International media have represented Yolanda, or what they call Haiyan, as an "enemy." But the intensity of this typhoon certainly was propelled by economic and political processes. One way to understand the impact of climate crises is by looking at it from an integrated disciplinal perspective that re-centres the local narratives of people who consistently experience these crises. Strong weather events like Yolanda are the result of too much toxification of our earth. The waters in the Pacific Ocean are warming. The temperature is rising due to the copious greenhouse gas emissions that this ocean has forcibly been swallowing up. One elder in Samar that I talked to said, "When the water is warm, it attracts huge typhoons." He is aware of his science! The warming of water is perpetrated by the pollution that is being generated by highly industrialized countries. We should ask how much greenhouse gas emission is being contributed by little island communities in the Philippines, or rural villages in Vietnam in the global warming. There is very, very little of that. But the irony is that these countries receive the biggest impact of climate change and climate crises. This is how climate injustice works in countries like the Philippines: it results in the devastation of local communities where lives are tragically taken due to the intense weather events that strike these places. Extreme consumption reinforced by neoliberalism is embedded in the issues around climate change and climate justice. For example, there are cases of mineral extraction happening in the Philippines by multinational corporations from highly industrialized countries, and we know that industrialization contributes to greenhouse gas emissions. Here, I would like to emphasize how the continuing colonialism perpetuates the destructions brought by strong weather events. Within this complexity, I reflect on this question: how do Waray elders understand typhoons, and what is the local knowledge around ecological stewardship? For me, the local community members must be given a voice to express their own experiences in tackling climate crises. Their voice is important in my research as I want to hear the local elders and community members who have direct experiences of the crises brought by climate change (Figure 3.1).

KIRSTEN Let's hold that thought—we will be discussing this soon. Lara, I want to hear your research question.

Figure 3.1 Public school teachers from the Municipality of Guiuan perform their devised community-based theatre piece on Super Typhoon Yolanda among the community members of Barangay San Pedro Tubabao Island. This performance was the final output of a week-long applied theatre workshop that mobilized the participation of local government officials, school administrators, elders and artists. (Photo by Dennis D. Gupa.)

LARA So, yes, going back to my question: I am mainly interested in bringing together local traditional knowledge and scientific knowledge through theatre to further understand and strengthen how we adapt to our catastrophic futures. And my research question is: "How applied theatre can, as a medium for dialogue, build collaboration between scientific and traditional ecological knowledge for co-creating strategies on climate crisis and adaptation? Especially in relation to what we value, and how we are adapting to climate change. How can art become a tool to facilitate these spaces of adaptation?" These are my broad questions. And during my fieldwork I am planning to facilitate an arts-based process where knowledge holders are going to share their perspective, relations and experiences on the local and global climate crises.

KIRSTEN What I hear from both of you is that you are looking for a way that we can co-create knowledge: for example, scientific knowledge and Indigenous knowledge. I would be interested to know how, where and when are we thinking about co-creation of knowledge. You are both using co-creation more as a methodology?

DENNIS I believe so. For me applied theatre is a method of enquiry in which the acts of co-creation are a significant part of this enquiry; mobilizing different voices in rendering other possible futures. I think that applied theatre thrives with a collaborative effort coming from the participants. While I was at my field site, I noticed that I was seeing the world through the lens of applied theatre, and it allowed me to see what the local artistic creation practices were. This community is surrounded by water. The sense of fluidity informs people's ways of knowing. Their lives flow, interact and overlap with the ocean every day, from subsistence to annual religious commemorations of their patron saint, and the history of their communal belonging revolves around the water. Against this backdrop, it is extremely problematic to impose a linear and hierarchical conceptualization of knowledge production. The ontology of applied theatre is emancipatory. I found Sally Mackey's "fluidity of epistemology" (478) and Dwight Conquergood's "dialogical performance" (9) helpful in my conception of an arts-based participatory co-creation of knowledge in this community. Mackey suggests the metaphor of "polyphonic conversations" (478–480) that can challenge the structures of power embedded in knowledge production, by allowing various voices to surface in these conversations. This is a relevant insight in my work within zones of precarity where community members are not given a considerable amount of opportunities to articulate their narratives. I think that the metaphor of many voices resonates with Conquergood's "dialogical performance" that engages a continuum of conversation, and that restores an open dialogue that is emancipatory and inclusive; as he said, "[i]t is a kind of performance that resists conclusion" (9). Conquergood quotes Henry Glassie in his work, borrowing his notion of "intimate conversation" (10), a concept that helped me to conduct my daily life ethically on the island while engaging with the people there. That is deep. For me it means living in the field site and being with the people within their natural and local ecosphere. I owe a lot to Mackey's and Conquergood's wisdom. The durational movement of the tides in the island is telling of its life. Listening to the stories of people in the community is humbling. This "intimate conversation" is similar as the Waray's *iru-istoryahan* or "story sharing." When a group of people gathers in front of the house you would see some snacks of boiled root crops or coconut wine which the locals call "tuba." This gathering with food allows private stories to be set free. And in the process of sharing their stories, their sense of making their world is shared and felt; *naábat,* as they call it. I have joined this story-sharing and I have noticed the performative aspect of it. Sometimes they have fun with each other during Videoke sessions inside their houses, or in their backyards. One time I heard jokes about the Super Typhoon Yolanda; it was hilarious. This is a people who can still make fun of their tragedy. At times the stakes are high, especially when men are drunk and hot issues are tackled, like religion. Joining this *ero-estoryahan* is an intimate engagement with the people and with their ways of knowing. So, when I endeavoured to create a performance with the community members, I integrated this process of *ero-estoryahan*. This familiar form of communication offers us an understanding of what we wish to create. I think the process of co-creation of knowledge must begin from the people within their natural environment, and from their local culture. Their voices must be at the centre of participatory theatre and community-based-theatre performances.

KIRSTEN I would be really interested to know, Lara, how you relate this intimate art form to probably the main theme of your research, climate justice. You are talking about climate justice, climate adaptation: where and how does applied theatre come in?

LARA I think it is important to note that at this moment in time many actions and small revolutions are emerging, which is a wonderful, joyful thing. It resembles what Bogad calls "tactical carnival," a social change of the activist, the unpredictable and participatory actions that can create a joyful counterculture (57). We imagine what climate justice is and create the world we want to live in; we occupy public spaces and we reclaim our voices to transform these moments through participatory action. We practise and live the utopia we imagine, which is a radical manifestation and a social critique of our societies, in action. So, as a way of re-imagining and co-creating,[3] we are simultaneously learning from and teaching each other to collectively rehearse and build our future in the present, by means of performative pedagogies. I think these steps are all interrelated with my understanding of applied theatre. Thus, as Denzin puts it, "performance through critical pedagogy disrupts those hegemonic cultural and educational practices that reproduce the logics of neoliberal conservatism" (383). Going back to what Dennis mentioned earlier about Mackey's fluid epistemology and Conquergood's "dialogical performance," I want to add that within climate justice conversations, we need to try to grasp the plural nature of this issue. Because we are facing an entangled, "wicked problem,"[4] that does not come with a solution; there is no formula, no singular definition or guide map, no true or false; it is impossible to see it from one perspective, one theory, one worldview. We need to accept the plurality, the chaos, and accept that this is not something for which you can find a solution; it is complex like a living organism, and we need to equip ourselves to see it from that perspective of cycles; to observe, reflect, take action and repeat. It is difficult to practise this since we are surrounded by Western epistemologies and ontologies that try to oversimplify the problem by deconstructing it to its cells to find a solution. But it ignores the complexity of the whole picture; the relations, the interconnectedness and the behaviour of the problem. So, many people are coming to understand that the monocentric idea of knowledge might not be the only way of approaching the climate crisis, but that it needs the connection between art and other ontologies and epistemologies. I see art and applied theatre as extremely powerful instruments that we could bring into the conversation when we do not have common languages, but we have common struggles. I think applied theatre could become our common language.

KIRSTEN The big question now is applied theatre. Anton Chekhov argued that theatre does not solve the problems, but accurately describes the issues at stake. We come up with the problems and theorize them. But we rarely find immediate solutions. As Lara said, I am not asking you to come up with a solution for this major issue. But what I'm interested in is to explore the ways creative outputs contribute to potential solutions for major political, social and environmental issues. What can we do to really start probing the community of stakeholders for the actions that are required? What should we do now? Now that we have looked at this problem, what are the realistic next steps that we can take in our field?

(From the perspective of an applied theatre scholar and reflecting on the years of teaching in a variety of contexts, I am acutely aware that our creative outputs only bring change once other political, social and economical systems are adjusted, replaced or even removed. As Michael Balfour mentions, the "little changes" [348] we can make in our work should be our starting point, and therefore I believe we should focus on the realistic changes we can make as educators: training our students to become ethical, professional facilitators who are capable of creating a playful and safe space for their participants to discuss their needs and challenges while enjoying the beauty of our incredible art form. Facilitation is a complex skill set that requires extensive training hours and a considerable amount of reflection. Additionally, applied theatre practitioners are artists and should not lose sight of their aesthetic practice. Therefore, part of applied theatre training should focus on the artistry and craft. It will strengthen our field. I asked Dennis and Lara the same question).

DENNIS Creative outputs are helpful, only if the process is empowering. We should know our craft and the theories behind it. Because as applied theatre artists, scholars and teachers, we are tasked to reflect and act amidst the complexity of things when we are executing our projects. What I learned from my field research is how to stay patient in tackling the complexity of issues around climate crises. But at the same time, we also need to be responsive to moving the issues forward so that the institutions will not forget about the meanings and goals of our work. In an island community that does not have basic services, like access to clean drinking water, proper land and sea transportation systems, a dearth of health programmes for children, who look undernourished, and the depletion of marine resources that continues to alarm the fishers, the art that we make should directly matter to people.

I lived in the island community for two summers, and I was confronted by these facts of life. Many people in this island are forced to live in the darkness of economic scarcity and political debacles. How does theatre contribute to making things better? I needed this interrogation to go beyond what we in the Philippines usually think of theatre as a form of entertainment. For me, theatre is a social practice and it should be employed to examine the suffering of my community, and as an artist I do not want to divorce myself from carrying the burden of people who are suffering. If the national government fails to perform its functions in solving local issues, then we can innovate ways for social cohesion to happen through the arts. For example, I co-organized a theatre festival and conference while I was conducting my field research. It was attended by community members, artists, scholars, students, policymakers and government officials. Theatre can create possibilities where intercultural and intersectoral dialogues can take place between artists and people who are directly responsible for crafting social policies. I think that is the answer to solving the issue; at least one way of solving the issue. The Waray have this practice of working together which they call *pintakasi*. It encourages community members to engage in a concerted effort that invokes *gu-ol*, or shared obligation in mobilizing community building. Here I would like to reflect on the etymology of "act," whose Latin root "actus," means "to do, to perform actions, to achieve, to accomplish, to take action, to do things" ("Act"). There is a past tense in this word. Even if you have not executed the work, it has already been done. I have learned how to navigate around bureaucratic processes that slowed down my research by including the local government officials in my creative process. I co-created community-based theatre performances that were presented to the community with the participation of local government officials. Usually in local theatre performances they are "relegated" to being audience members, but in the project that I co-created with the community members these local government officials willingly joined us as organizers of our theatrical productions. Maybe next time, they will find time to be our performers. We have to make them creatively and ethically invested in our work. These performances demonstrated the creative capacities of the elders and community members in articulating their stories about climate change. It's a two-fold objective: one is the gesture of transgressing those constructed social identities that perpetuate marginalization, and the other is to mobilize artistic projects where people can come together to discuss social issues without fear of being killed. Creating art is also about creating scenarios for community collaborators to experience agency, which means allowing people to be active in formulating their future. I see applied theatre as a way to respond to what bell hooks urges us to do, "pushing against oppressive boundaries set by race, sex, and class domination" (15), and to engage our "capacity to envision new, alternative, oppositional aesthetic acts" (ibid.). Such powerful words. I think we can obtain these "oppositional acts" by means of collaboration and by deploying new approaches to alliance-building (Figure 3.2).

I remember you, Kirsten, telling me before I went back to the Philippines to do my field research: you told me to drop my value

Figure 3.2 Outdoor performance of "Murupuro/Island of Constellation 2," devised by teachers from different public elementary schools in Guiuan and fishers of Barangay San Pedro. The workshop and final performance were facilitated by local artists of Sirang Theatre Ensemble (Tacloban), led by Amado Arjay Babon. (Photo by Aivee C. Badulid.)

judgements, however difficult and challenging; to engage with local government systems, and I felt delighted to receive the generosity of the individual staff members of these government institutions! They can be our collaborators to solve issues that they deem problematic. They know better about their community and their work than I do. Hopefully, the national government will become more caring when they hear about and witness our works with community members.

LARA Well, I will say no, applied theatre is not the only solution, but it is a great way to start asking more questions. The answer lies in taking action from multiple directions; without that action, we will be lost, we will not even be able to explore the problem in depth. So, I think the key is building community engagement through collaborative applied theatre processes, as solid pedagogical steps to approaching the problem. As I said before, we need to shift our understanding from a rigid, solution-oriented perspective to process-based encounters. I think there is something inspiring in knowing that you do not have an option to be passive during times of change, because as Greta says, "change is coming whether you like it or not" (Thunberg). Every choice you make is determining your future: today, what position are you going to take? What is going to be the process during and after the storm? This will determine our future. I bring it back to the urgency of responding to these questions now, as we are running out of time; we need to work as hard as we can *now*. More than ever, we need to practise, transform our realities and try to take care of each other to be able to work tirelessly. The world is going through catastrophes and experiencing the devastation of climate crises in different degrees. Even though we do not see the climate crisis going away anytime soon, it is crucial to engage in, as the world we know is collapsing, and how are we going to rebuild the ruins? People are starting to make the effort to create transdisciplinary dialogues to ask similar questions and search for responses through participation, collaboration and creativity. I see applied theatre as an exploratory tool to create a collective dramaturgy as a new language. I am trying to learn the ways we can start to speak this new language because we are creating a new language. I think people are striving to build connections with their communities and to collaborate with others to take an active role in climate action. It is beautiful to be a part of this, because through these collaborations we are in the process of re-defining what we know and how we can learn a new language. This action is a living organism, actively rebuilding itself, informing itself through a constant movement.

Dr. Woodrow "Woody" Morrison, a respected Haida elder, has taught me about the meaning of balance. It is essential to understand the balance within ourselves and in between each other. We need to think about the ways we can maintain balance, and I am trying to learn how to do this by healing myself and my engagements from Eurocentric epistemologies and ontologies. I am trying to detach myself from those monocentric worldviews and systems. We need to break this dynamic, the destructive dynamic that we live in today. I am thankful to Indigenous peoples, whose knowledge and stories stretch to time immemorial, and in between genocides. Indigenous warriors and elders do not give up on us and continue to guide us and generously share their knowledge with us; they are telling us how to maintain the balance through their stories, ceremonies and songs; we need to pay attention to these land defenders, political and performance pedagogies, and try to hear what they are saying to be able to heal.

KIRSTEN One last question. I'm going to ask you to think about why applied theatre is useful, helpful, compelling to when we talk about Indigenous ways of knowing. And in this context, we are still talking about climate justice. I'm going to ask you for one word only. The relationship between the words might connect us to Indigenous ways of knowing. See if we can just go back and forth one word at a time.

LARA Engagement.

DENNIS Plurality.

LARA Respect.

DENNIS Agency.

LARA Autonomy.

DENNIS Enduring, enduring love—please take that as one word.

LARA Healing.
KIRSTEN Let's stop there. Healing.

Notes

1 I use the word "data" to refer to the stories, performance texts, songs and interview materials that I obtained in my field research. I put this in single quotation marks as a marker of new ways of looking at these materials; not just as objects that need to be interpreted or examined, but as sources of materiality that suggests relationality and communality.
2 Please see: "Strategies for Climate Crisis Adaptation: Bringing Indigenous and Western Knowledge Systems together through Theatre."
3 In her article "*A good day out: Applied theatre, relationality and participation,*" Helen Nicholson calls this process "the prefixes that define the contemporary moment" (252).
4 Rittel, Horst W. J., and Melvin M. Webber. Dilemmas in a general theory of planning. *Policy Sciences, 4*(2), 1973, pp. 155–169.

Works Cited

"Act." *Oxford English Dictionary,* 2020. https://www.lexico.com; https://www.lexico.com/definition/act (accessed 19 August 2020).

Balfour, Michael. "The Politics of Intention: Looking for a Theatre of Little Changes." *Research and Drama Education the Journal of Applied Theatre and Performance,* vol. 14, no. 3, 2009, pp. 347–359.

Bogad, Larry M. "Tactical Carnival: Social Movements, Demonstrations, and Dialogical Performance." In *A Boal Companion: Dialogues on Theatre and Cultural Politics,* edited by Jan Cohen Cruz and Mady Schutzman, Routledge, 2006, pp. 46–58.

Conquergood, Dwight. *Performing as a Moral Act: Ethical Dimensions of the Ethnography of Performance.* University of Michigan Press, 2013. doi:10.3998/mpub.4845471.9

Denzin, Norman K. "Critical Pedagogy and Democratic Life or a Radical Democratic Pedagogy." *Cultural Studies Critical Methodologies,* vol. 9, no. 3, 2009, pp. 379–397.

hooks, bel. "Choosing the Margin as a Space of Radical Openness." *Framework: The Journal of Cinema and Media,* vol. 36, 1989, pp. 15–23.

Mackey, Sally. "Applied Theatre and Practice as Research: Polyphonic Conversations." *Research in Drama Education: The Journal of Applied Theatre and Performance,* vol. 21, no. 4, 2016, pp. 478–491. doi:10.1080/13569783.2016.1220250

Morrison, Woody. Personal interview. 25 February 2020.

Thunberg, Greta. "Watch Greta Thunberg's Impassioned Speech: 'Change is Coming Whether You Like it or Not.'" *NBC News,* 2019. www.youtube.com. (3:44). (accessed 23 September 2019).

4 Voicing students' perspectives in the transformation of theatre pedagogy for climate justice

Alexandra (Sasha) Kovacs

"What might we teach Swedish student activist Greta Thunberg if she were to choose post-secondary education in theatre?" When I first encountered this inquiry, a prompt for this collection, I was immediately struck by its paradoxical nature. While Thunberg has shaped the perspectives of countless youth, establishing a performance repertoire for climate activism insofar as her "gestures and behaviours [are those] that get re-enacted or reactivated again and again" (Taylor 10), she also has accelerated a distrust in the ability of and potential for educational institutions to incite real change. As noted in the 2019 *Time* magazine issue that named Thunberg person of the year, "hundreds of thousands of teenage 'Gretas,' from Lebanon to Liberia have skipped school to lead their peers in climate strikes around the world" (Alter et al., n.p.). It is outside the classroom that Thunberg positions the youth-driven climate movement's force and impact. Given this context, my response to the collection editors' provocation begins from an impulse to flip the question; to ask not what theatre educators might teach young activists if they pursued post-secondary theatre education, but instead what young activists recently graduated from theatre training environments can teach theatre educators about instructing *them*. Especially since recently graduated, current and incoming cohorts of students will "suffer more than 80% of the illnesses, injuries, and deaths attributable to climate change" (Sanson et al. 202), listening to and acknowledging students' experience and perspectives is key to developing an effective theatre pedagogy that engages students in a discussion of climate justice and its intersecting "social challenges" (Mason and Rigg vii). Inspired by James Engell's principle of "reciprocal mentorship" that acknowledges the "magnanimous dedication of teaching and learning from each other" (27) required of humanities education that aspires to engage with the climate crisis, this contribution queries the potential for theatre training to engage with climate justice by drawing on students' knowledge and perspectives. Positioning a select group of student voices alongside existing research into theatre pedagogy, I propose that theatre education's meaningful engagement with climate crisis would necessitate a divestment from current approaches to on-campus theatre production—which tend to perpetuate neoliberal ideals— and an embrace of alternative models that foster interdisciplinary and

intergenerational collaboration. Though the perspectives shared in this contribution are specific to the local context in Victoria, British Columbia, I expect the identified challenges and suggested approaches might be relevant for other post-secondary theatre training environments, where faculty may desire to establish a pedagogy that is responsive to the world's most pressing crisis.

Mapping climate activism at the University of Victoria

In 2018, I began teaching at the University of Victoria, on the lands of the Lekwungen-speaking (Songhees and Esquimalt Nations) and W̱SÁNEĆ peoples. During my first day on campus I attended a new faculty orientation event where the University emphasized its commitment to sustainability by sharing high gloss copies of its Strategic Framework (2018–2023) ("A Strategic Framework"). Of the six priorities listed in that organizational plan, the fifth includes the directive to "Promote Sustainable Futures." The university's vision for this pillar is explained in the document as follows: "The University of Victoria will be a global leader in environmental, social and institutional sustainability through our research, academic programs, campus operations, and the impact and influence of our students, faculty, staff and alumni" ("A Strategic Framework" 7). The achievement of that aspiration is central to much of the University's public messaging. The University of Victoria celebrates its "work shaping public policy discussions and climate responses, and taking practical steps towards low-carbon energy solutions" ("UVic Ranked as Global Leader"). Most recently, in April 2020, my inbox was flooded with the news of the Times Higher Education global impact ranking that placed UVic fourth among the world's top-tier leaders in climate action (following behind the leading institutions of University of British Columbia, Vrije Universiteit Amsterdam and University of Tasmania). This ranking also placed UVic in the world's top 100 institutions overall in contributing to a sustainable future for the planet ("UVic ranked as a global leader"). Since my first day on the job, I have seen the University of Victoria proudly perform its position as a recognized leader in climate change action.

However, it is students who consistently hold the University accountable to its sustainability commitments. In September 2019, the University of Victoria Student Society (UVSS) released a statement pointing to an increasing disjunction between the rhetoric of the University's sustainability signalling and its actual commitments. In a press release that endorsed the actions of a group called "Divest UVIC," who organized protests over the UVic Board of Governors' climate inaction, the student society stated: "Students at UVic join increasing global demands for action on climate change," and went on to note that "UVic has ignored calls to divest from fossil fuels from students, staff, and alumni since 2013" ("UVIC Student Society"). In response, the students coordinated a mass walkout to "demonstrate to the University that students will not stand idly by as they profit from the destruction of the planet" (Kozelj). This context at the University of Victoria supports recent research which demonstrates how "students continue to be at the helm of a growing social movement to demand that academic institutions take action on climate change and divest their endowments of holdings in the fossil fuel industry" (Grady-Benson and Sarathy 662).

Yet while student groups at the University of Victoria might organize outside the classroom to engage in climate justice activism (Bennett i), within the Department of Theatre there is little evidence of such dedication, especially with respect to on-campus production. Like many other theatre departments across North America, our Department's undergraduate curriculum is interwoven with its production of a mainstage performance season, called the "Phoenix Theatre." Since starting my position at the University, there has not been one mainstage production that has engaged climate justice as a theme or practice ("Season Archive"). The dearth of creative activity around climate justice is not isolated to this particular department and its educational context of creative production; as early as 2005, environmental studies scholar Bill McKibben was asking why cultural activity about climate crisis is so non-existent:

> One species, ours, has by itself in the course of a couple of generations managed to powerfully raise the temperature of an entire planet, to knock its most basic systems out of kilter. But oddly, though we know about it, we don't *know* about it. It hasn't registered in our gut; it isn't part of our culture. Where are the books? The poems? The plays? The goddamn operas? Compare it to, say, the horror of AIDS in the last two decades, which has produced a staggering outpouring of art that, in turn, has had real political effect. I mean, when people someday look back on our moment, the single most significant item will doubtless be the sudden spiking temperature. But they'll have a hell of a time figuring out what it meant to us.
>
> [emphasis in original]

While a quick overview of the department's current production activity might lead one to conclude that the climate crisis is not, to echo McKibben, part of our departmental culture, this would neglect the continual engagement I have experienced with students—inside and outside the classroom—in discussions of the climate crisis. If students are not taking up these considerations through the department's venues and performance opportunities, are they doing so elsewhere? If so, why? How can my own theatre department offer students opportunities to grapple with this global issue that is so essential to their mental and physical survival?

Where is the theatre about climate justice? Listening to the students' voice

I turn to the voices of students themselves to explore these challenging questions. A growing body of research in student voice initiatives emphasizes the significance of "a transformative, 'transversal' approach" to education "in which the voices of students, teachers and significant others involved in the process of education construct ways of working that are emancipatory in both process and outcome" (Fielding 124). These student voice initiatives set out to adjust and revise educational practices by acknowledging "the experiences, views and accounts of young people" (Bourke and Loveridge 2). Such a centring of the student voice is recognized for its potential to "sensitize our research efforts as well as inform and strengthen social justice leadership and transformative learning spaces" (Mansfield 393). Guided by this research, I conducted interviews with a number of recent alumni to ask what aspects of their theatre training supported or inhibited their exploration of climate justice and its intersecting concerns. In total, I spoke to eight students who graduated within the last two years, each in interviews of approximately sixty to ninety minutes. This small group of students was selected for interview because of their experience engaging with climate or social justice issues during their theatre training. Their perspectives are in no way representative of all student experience in the department. On the contrary, the students interviewed occupy a minority position in that they are some of the few that took up, during their theatre training, issues of climate and social justice. Yet within their interviews, these students surfaced critical concerns about the structures of theatre production and curriculum that foster or impede explorations of the intersections between theatre and climate justice.

The most striking discussion, persistent across all student interviews, concerned the prioritization of the department's mainstage season. Students characterized this season as the "core" curricular site that encouraged important vocational understanding (stage-craft learning experience), but took up space and resources that might otherwise have allowed for their investigation of contemporary social, political or cultural issues. Language of consumerism, consumption and resource depletion characterized student discussion of the season. A recently graduated applied theatre student, who worked in the theatre box office during the course of her degree, considered how even the architecture of the building emphasized the Department's prioritization of the mainstage curriculum. She notes that "when you walk into the building the box office is the first thing you see on your right" (EB). Of course, this architectural condition serves practical purposes—it makes the purchasing of tickets to student performances readily accessible to the public—but it also speaks symbolically to a consumer-driven priority in the very design of the theatre education. Across many interviews, students discussed how throughout their education they were consistently made "aware of the money that [they're] bringing in and the tickets that [they're] selling and the subscriptions" (EB). Students discussed their consciousness of the season's box-office sales, describing how his "university education kind of flows around that model of production and presentation and also box office, which becomes a key component of the cycle," and that this places students on a "presentation house train

... [that] doesn't necessarily offer tools for students to consider how to use their artistic work as a form of activism" (KD). Production was discussed in the interviews as paramount to the department's activities, one student noting that the "theatre department is designed to produce and produce and produce and get subscribers" in a kind of "fast food" model (KD). Such a perspective is paralleled by another student who recalls, "I think the Phoenix was often referred to as like a machine—it produces the mainstages, and it produces these things in this order and it's a well-oiled machine" (MP). Attached to this discussion of the Phoenix theatre as a factory, students highlighted the unsustainability of serving and labouring for this model, suggesting that the "theatre culture in our department definitely was encouraging overwork. I mean, you couldn't survive a semester without doing a couple of all-nighters at least, like it just wasn't possible without that. Maybe they weren't encouraging that but it wasn't possible" (LA). The burnout culture attached to the mainstage was also identified as the particular aspect that discouraged students from building connections between their practice and their political commitments: "Everyone was like, 'oh, we're too tired to come after rehearsal to the legislature to go protest something … We're a little too tired after rehearsal'" (MP). The mainstage was continually described as a set of industrial practices that depleted students' ability to align their work or energy with activist commitments.

These shared experiences recall a structure for theatre pedagogy that Jill Dolan critiques in her 2001 book *Geographies of Learning* as "succumb[ing] to marketplace pressures" (51). The mainstage production model's operation as a factory that panders to a particular perception of audience desires (ibid.) is, as Dolan suggests, an impediment to the realization of theatre's pedagogical potential. It is also a site that reinforces neoliberal doctrine. In her introduction to *Theatre and Performance in the Neoliberal Academy,* Kim Solga suggests that neoliberalism's entrenchment in the university is evidenced by the increasing emphasis on "skills-training aimed at individual student consumers," which breaks down the potential for university education to develop a "community of concerned citizens" (5). Overwhelmingly, the student interviews emphasize that the mainstage aspect of the curriculum, experienced by students as a pivotal component of the programme, reinforced the key "governing social and political rationality" of neoliberalism "that submits all human activities, values, institutions and practices to market principles" (Brown qtd. in Solga 5). This is a function of the production model operating within the larger corporate structure of the University of Victoria— the mainstage operates as a public–private partnership, in that its budgets are dependent primarily on box-office revenue generated by community subscribers who the department often presumes expect a performance product that models mainstream regional theatre. The challenge for any attempt to transform this model is that these mainstage productions are what draw many students to engage with theatre education in the first place. In a promotional recruitment video, an undergraduate production student offers enthusiastically that "the department runs as an operating theatre company" ("UVic"). For many students, whose views are not represented in the interviews I conducted, a skills-based experiential pedagogy that mimics industry-producing realities is attractive. Jill Dolan acknowledged the challenge of divesting from pedagogies that allow students to experience and work through the economic realities of surviving as a theatre artist, saying "it is excessively privileged to suggest that an arts education is more important than a livelihood" (52). Of course, theatre educators must acknowledge our obligation to offer students the skills to survive in a challenging industry. However, we must also consider our moral imperative to provide opportunities for students to activate theatre as a tool to advocate for the planet's survival.

While the mainstage's neoliberal model was defined by students' perspectives as incompatible with climate justice activism, courses in applied theatre and theatre history were identified as contexts that provided opportunities to develop knowledge about theatre's activist potential. Students spoke about applied theatre courses as "the foundation for further and future projects" that addressed activism and helped students "find ways to use research as a tool in theatre" (SH). Students identified their professors working in this area—Dr. Sadeghi-Yekta, another contributor included in this volume, Dr. Warwick Dobson and Dr. Yasmine Kandil—as key mentors and guides in this work. However, students also recognized "a hierarchy within the department where applied theatre was kind of at the bottom" (EB) and noted that this devaluing of applied theatre practice is most apparent in the lack of space given to these projects in mainstage programming. Theatre history courses were, like applied theatre, described as productive sites for learning about "theatre being used as a tool for democracy and other political change" (LF). Learning in theatre history courses was often discussed as the motivating force for performances that engaged with climate justice and its intersecting activisms (KD; LA). As a theatre historian myself, this acknowledgement of theatre history's potential to incubate the development of students' techniques in activist performance was exciting and energizing, especially because it runs contrary to the extant literature, which seems to predict student disengagement with such education in the discipline. In a 1994 issue of *Theatre Topics* devoted to the subject of theatre history, Jerry Dickey and Julie Lee Olivia note that "undergraduate theatre students typically approach their required courses in the history of their art with apprehension at best and downright dread at worst" (45). They cite the work of the pioneering historian Oscar Brockett, who in 1988 bemoaned student motivation in theatre history courses as follows: "first of all, to pass the course and, only secondarily, to learn something about the subject" (46). The perspectives shared across the interviews I conducted with these students served to resist these perceptions, and instead parallel the offering made by scholar-practitioner Lissa Tyler Renaud who considers the importance of intellectual life within theatre training, noting that "if our goal is to foster meaningful inquiry into the theatre as a whole, we have to give our students the intellectual tools to make the inquiry" (86).

But while theatre history and applied theatre approaches were identified by students as key to appreciating the mobilization of theatre for environmental and social justice purposes, the students consistently articulated dissatisfaction at the disjunction between classroom learning and departmental production practice. As one student remarked, "you'd never see the same activist play [that was read in a theatre history class] being done in the department" (EW). The lack of alignment between in-classroom learning and experiential creative practice was consistently discussed as a frustrating dichotomy:

> I would like to see a little bit more 'practise what you preach.' A lot of stuff that we learn in theatre history I find fascinating because it is so subversive or controversial but then the theatre [produced by the department] shies away from that stuff when they put stuff onto their mainstage for the public. I understand this might be because of money and stuff like that. But I just think that what goes on the stage should maybe also reflect the stuff that you're talking about in the classroom. Because it is a post-secondary institution and it is not a business. And these are actually students and they are doing this for their education.
>
> (EW)

This same student remarked how one of their most memorable experiences across their theatre history courses was conducting a collaborative research project on the topic of climate change in theatre. Through that project the students collectively "did a ton of research on different universities that have policies on environmental change" and considered how that could be applied to the department's processes for the mainstage season. However, the student also noted that this kind of learning was not integrated into any outcome; "it was just a student project and nothing got applied" (EW). Interviews stressed that although theatre history and applied theatre courses offer students the tools for activist-informed performance, there is little opportunity to activate this learning through the department's experiential programming—its mainstage—viewed by these students as the major indicator of the department's curricular values.

Given this environment, students pursued practice that addressed climate and social justice through: a) student-led productions; b) directed study courses; or c) course projects that allowed for guided independent creative practice. In all these contexts, students consistently discussed the challenges they experienced gaining credit and finding resources for this work. The Student Alternative Theatre Company (SATCo) was identified by students as the only venue that produced performances that "overtly spoke to climate crisis" (MP; TN). This production context developed and supported the creation of devised theatre projects like *Waste Management* and *Are We All Dead*. The co-creators of the latter performance were among the students interviewed, and their reflections on the performance reveal the limitations inherent in the practices of this student company:

> *Are We All Dead* was trying to resurrect and refurbish agitprop strate-
> gies to address the climate crisis. … We were equipped with resources
> through the Student Alternative Theatre Company, which is granted
> rehearsal time and stage time and space to present student-written,
> -directed, -designed and -performed works. The SATCo structure is vol-
> unteer—we never received any sort of credit for this production, and
> all rehearsals and work for the show took place in the hours outside of
> the course curriculum. In total we had thirty or forty hours of rehearsal
> over a three-week period to construct a thirty- or forty- minute piece
> that was devised.
>
> (KD)

Outside the auspices of SATCo, students also developed environmental
and social justice performances through directed studies, but the success
of these projects was dependent upon a student's ability to secure resources
outside the department. One student described the difficulty of accessing
departmental resources for a devised project as part of an independent
study credit:

> Booking a space was so difficult. I couldn't book a space. Financially,
> I had to do it all myself. I ended up putting up close to $900 outside
> of course payment. I think that was the only time I've ever cried in a
> professor's office.
>
> (MP)

This student described their relief at eventually finding a venue off campus
to present their work. Students often spoke of the need to access outside
facilities to present work because of limited access to space and resources,
due in part to the requirements of the mainstage production model that
makes rehearsal space scarce. Such reflections compel the question: how
can my own theatre department transform its practices to support these
student-led explorations of the world's most pressing global issue?

Proposals for transformation

These student interviews identify two areas—mainstage productions and
curriculum—that need to be restructured in order for theatre education to
address the marginalization of climate activism projects and pursuits within
the Department of Theatre at the University of Victoria. In terms of the
approach to mainstage production, one recent performance alumna suggests:

> I do think within our mainstages, we could talk more explicitly about
> these motivations for these shows. And rather than just presenting the
> audience with something, we could also start a dialogue so that the
> learning goes beyond the aesthetics of the production.
>
> (TN)

This perspective echoes the viewpoints expressed in a 2004 *Theatre Survey* article by W.B. Worthen, who asks: "might the season, or some element of the season, also be understood in a different sense, as a kind of experiment, as a kind of inquiry?" (266). The alumna's perspective also aligns with the suggestions offered in Jill Dolan's "Road Map toward Addressing the Theory/Practice Split in Theater Studies" that calls for a recognition of "theater as pedagogy" (63). Offering discursive spaces (talkbacks, lectures, workshops) to reflect upon campus production through an ecocritical lens might be a good start here; but in order for theatre to develop a pedagogy that can intervene to offer more meaningfully integrated teaching and action around climate justice, more work also needs to be done to transform the overall approach to play selection, development and programming.

Like many other departments across North America, the present mainstage model of UVic Department of Theatre "apes the structure of regional theatres around the country" (Dolan 51) and typically produces canonical plays that can guarantee box-office success. Research confirms how this approach to on-campus theatre production, in the Canadian context, "entrench[es] mindsets and perspectives that stay with graduates throughout their professional lives" (Hanson and Elser 36). More recently, in response to the recognition of on-campus theatre production's perpetuation of gender inequities, initiatives like The Pledge Project (Production Listing to Enhance Diversity and Gender Equity) were established to "help educators address the lack of gender equity in post-secondary theatre programs." This project features a fully searchable database with a listing of plays written by Canadian women with a cast size of six or more. The database is intended for theatre training and educational settings where large cast sizes are needed to accommodate student participation. The project also "invites schools, departments, professors, visiting directors, and the like to make public pledges, that they post on their website, to improve the representation of women playwrights and other marginalized communities at their institutions" ("About"). These initiatives that redress gender imbalance in on-campus production offer models that could be generative for the project to further engage mainstage productions in deeper conversation with climate justice. Developing a database of plays that engage with themes or topics concerning the climate crisis, and that could be produced within a mainstage season, would be one way to potentially increase the engagement of department with pedagogies of climate justice. Playwright and co-founder of Climate Change Theatre Action (CCTA) Chantal Bilodeau (whose colleague and co-founder Caridad Svich was interviewed for this volume—see chapter 10) offers a start to such an initiative with her "list of climate change plays" ("Creating a List of Climate Change Plays"). Available on the *Artists and Climate Change* initiative website, it also provides information on university courses and programmes that address the intersection of the arts and climate change.

While these resources offer pathways to engage climate justice within departmental theatre practice, it is wise to anticipate resistance to such shifts in approach. In many departmental meetings, I have heard proposals to engage pressing global issues in the season stymied because of a pre-sumed fear that such works would not be of interest to the current sub-scriber base. Again, the box office is the first thing driving these discussions. In response to this pressure, I suggest that departments must develop more creativity in their approach to building and fostering stakeholders and audiences; to shift campus production away from a neoliberal consumer model to one that privileges inquiry. We might draw on the strategies of the fossil-fuel divestment movement, which was "inspired by a powerful history of students and communities calling for institutional investments to match the values of those institutions" (Lenferna 139). Similarly, these student subjects call for a theatre pedagogy that aligns its resource use with the premise that "imaginative engagement in the arts provides real expe-riences that change who we are and can motivate progressive change in the world" (McConachie 98). More interdisciplinary and intergenerational collaboration in the development of mainstage programming will be key to divestment from market-driven approaches to on-campus production. We have only just begun this process at the University of Victoria. I am currently working in collaboration with an applied theatre colleague, Dr. Yasmine Kandil, to develop a three-year project that aims to restructure approaches to programming the mainstage season by inviting cross-campus and com-munity collaboration. By approaching mainstage production through an interdisciplinary and intergenerational committee—composed of stu-dent, inter-departmental faculty and community participants—we hope to revision the Phoenix theatre as a venue that operates in response to the concerns of the artistic and intellectual community in which we and our students live. In this model, we can invite representation from important university players in climate change research, such as the Pacific Institute for Climate Solutions, as well as student leaders who drive climate action; with these collaborators, we hope to jointly determine how best to trans-form the department's theatre venues into pedagogical spaces that serve a commitment to developing broader public knowledge, awareness and action, concerning climate change and other intersecting issues of equity.

In addition to inviting opportunities for cross-pollination, a theatre pedagogy for climate justice demands adjustments to curriculum that can compel students to practise and activate interdisciplinary thinking. One student interviewee discussed how interdisciplinary engagement was crit-ical to her environmental studies programme but was discouraged within the context of theatre:

> Looking back, I don't think that I would have been able to productively stick with only doing the theatre program without having my other major [environmental studies] there. I was always thinking about how the two were connected and I think that is really important to the-atre—practitioners need knowledge that is outside of the theatrical realm to connect their work to.

(LA)

The student went on to identify that their environmental studies programme requires the double-major structure,

> [b]ecause in that department, they recognize the interdisciplinary nature of it, like you have to be doing it with biology, or geography or political science or economics so that the environmental studies is informing that other part of your work and your life.
>
> (LA)

Across all the interviews, students described how projects that engaged with environmental/social justice were fostered through dynamic interdisciplinary alliances, through their work in environmental studies, gender and women's studies or sociology (LA; EB; LF). However, student responses emphasize that course requirements for their theatre focus areas often impeded their ability to develop interdisciplinary knowledge formation. In particular, the performance-focused (acting) students who were interviewed mentioned that the requirements for their programme prevented their engagement with applied theatre courses and other electives: "Our acting or performance curriculum is quite rigid in terms of the courses we have to take and so it didn't necessarily allow me to pursue the applied theater side of our program or department" (SH). Other students suggested that the heavy course requirements needed for their degree programme, coupled with extra-curricular theatre projects that were often non-credit, discouraged an engagement with global issues:

> If you're working too hard then you can't be in tune with the state of the world and global news and global politics. You don't have time to take a step back and read some good literature by Leanne [Betasamosake] Simpson. Without that you're not going to be making your best work.
>
> (LA)

Scholars like James Engell emphasize that a response to the climate crisis requires an "unprecedented coordination of science, politics, international relations, culture, technology, the arts, religious stewardship, and education" (24). Theatre is well positioned to contribute to this coordinated effort because collaboration is so key to its practice. Students recognize this potential: many of the interviewees spoke to the collaborative context of the theatre department as unique to their university learning experience, and for some the very impetus for their decision to major in the programme (SH; LA; EB). However, there needs to be a reconsideration of the curricular structures that impede the capacity of students to cooperate beyond their individual focus areas and to bring ideas developed from their engagement in other disciplines back into the theatre training environment. Let theatre educators divest themselves of the fear that a mastery of craft will be jeopardized by the gesture to open up time and space to encourage student engagement with other disciplines. Let us instead, as Lissa Tyler Renaud also advises, encourage an "interdisciplinary foundation" that can be mobilized through theatre practice (86). Engagement in this process shall not be "extra-curricular" or ancillary to a "foundational" set of courses, but a mandatory component of a theatre education that is responsive to climate crisis.

Conclusions: contending with a theatre in crisis

On 27 September 2019, Greta Thunberg spoke in Montreal at the Week for Future: Climate Strike. Addressing a crowd of young activists striking from school to march for a living planet and a safe future she said:

> [T]hey say we shouldn't worry, that we should look forward to a bright future. But they forget that if they would have done their job, we wouldn't need to worry. If they had started in time then the crisis would not be the crisis it is today. And we promise: once they start to do their job and take responsibility, we will stop worrying and go back to school, go back to work.
>
> (134)

The interviews I conducted in the course of this research make me reflect deeply on the current structures of theatre education within my own local context, which marginalize the students' perspectives that seek to engage with climate change and its intersection with social and political concerns. While these bright students committed to engage theatre as a tool to change the world, they did so on their own and without support or credit. When we can transform our classrooms—and stages—to become spaces that are responsive to student voices, we can revive the potential for theatre education to function as an agent for climate activism and positive change.

Now is the time to centre these student voices, if only because the present health crisis has paused theatre production in all theatre departments across the country, and indeed the world. On March 13, 2020 I took a break from composing this chapter to attend our department's production of Lillian Hellman's *The Children's Hour*. On way to the theatre I received a phone call informing me that the show was cancelled, shut down on the order of the Provincial Health Officer because of the risk of transmission of the COVID-19 virus. Since that March evening, the pandemic has reshaped human relationships, behaviours and working environments in ways I never thought possible. Many climate activists are now asking, in the wake of political leaders' urgent action and science-informed decision-making in response to COVID-19, why such a response was never mobilized for the more pressing and intersecting threat of global climate change (Andrews n.p.). We can only hope that "lessons Canadians are learning now in the fight against COVID-19 can be applied to efforts to tackle climate change" in the future (Andrews n.p.). Within the educational context the COVID-19 pandemic has led to overhauls of curriculum and demanded new approaches to instruction. In theatre education, we have had to exercise our creativity in unprecedented ways as our classrooms are shuttered and production opportunities cancelled until further notice. By May 2020 one headline in *The Atlantic* discussing the future of theatre read, "When Will We Want to Be in a Room Full of Strangers Again?" (Lewis). Though this pandemic threatens a bleak future for theatre, it also offers theatre educators a moment to strategize how they might dismantle a neoliberal pedagogy that risks swallowing up any potential for students to use their education to wrestle with climate justice. COVID-19's disruption of our habitual practices has the potential to catalyse theatre pedagogy's divestment from approaches to creative education that systemically marginalize and discourage learning regarding theatre's potential to inspire personal, political and collective change. Let theatre educators use this extraordinary time to continue collaborating with our students outside the classroom (as we are forced to do), in a spirit of reciprocal mentorship, to divest from neoliberal models of education that contribute to the climate crisis. In the process, we might devise a new role for theatre in the navigation of a most uncertain future.

Practically speaking …

- Map the geographies of activist performance at work in your own department. Can you identify the contexts in which students and faculty are addressing the climate crisis and its intersecting social justice movements through creative practice and research? Is this activity central or peripheral to your department's creative outputs? How can your department redistribute and reallocate resources to better centre this creative activity?
- Divest from a box-office driven approach to theatre production in the university. Instead of producing work according to certain expectations about audience desires, develop a working group with intergenerational and interdisciplinary membership that balances student, staff, faculty and community input to advise on the allocation of departmental resources for production. Reach out to climate scientists working in your university to contribute to this working group. Consider how your department's production activities can mobilize public awareness of scientific findings about climate change.
- Introduce on-campus production programming committees to Chantal Bilodeau's list. Develop that list by contributing plays and performances produced in your local context that address climate crisis.
- Make interdisciplinary learning a requirement for theatre majors. Advocate for curriculum changes that will encourage, and not penalize, student exploration of fields, practices and issues outside of discipline-specific theatre training.

- Offer students academic credit, mentorship, space and resources for devising original performances that engage with climate justice and its intersecting concerns.
- Cultivate an environment for student-driven creative practice that can harness the energy of youth to inspire change. Develop forums to engage students' perspectives regarding curriculum and practice. Take time to listen and respond to the student voices that consistently drive climate action and demand accountability.

Works cited

"A Strategic Framework for the University of Victoria: 2018–2023." 2020. https://www.uvic.ca/strategicframework/assets/docs/strategic-framework-2018.pdf (accessed 27 August 2020).

About. "*The Pledge Project.*" 2020. https://www.pledgeproject.ca/about/ (accessed 27 August 2020).

Alter, Charlotte, Suyin Haynes, and Justin Worland. "*TIME 2019 Person of the Year: Greta Thunberg.*" 2020. https://time.com/person-of-the-year-2019-greta-thunberg/ (accessed 27 August 2020).

Andrews, Alan. "*Canada's Coronavirus Response a Blueprint for Climate Action.*" *Ecojustice.* 21 April 2020. https://ecojustice.ca/canadas-covid-19-response-a-blueprint-for-climate-action/ (accessed 27 August 2020).

Bennett, Paige. "*University of Victoria Student Survey 2019: Climate Solutions Engagement Opportunities.*" *Pacific Institute for Climate Solutions.* April 2020. https://pics.uvic.ca/sites/default/files/PICSStudentSurvey_Report_finalApril22.pdf (accessed 27 August 2020).

Bilodeau, Chantal. "*Creating a List of Climate Change Plays.*" *Artists and Climate Change: Building Earth Connections.* 1 November 2014. https://artistsandclimatechange.com/2014/11/01/creating-a-list-of-climate-change-plays/ (accessed 27 August 2020).

Bourke, Roseanna, and Judith Loveridge. *Radical Collegiality through Student Voice: Educational Experience, Policy and Practice.* New York, Springer, 2018, doi:10.1007/978-981-13-1858-0.

Dickey, Jerry R. (Jerry Richard), and Judy Lee Oliva. "Multiplicity and Freedom in Theatre History Pedagogy: A Reassessment of the Undergraduate Survey Course." *Theatre Topics*, vol. 4, no. 1, 1994, pp. 45–58.

Dolan, Jill. *Geographies of Learning: Theory and Practice, Activism and Performance.* Middletown, Conn, Wesleyan University Press, 2001.

Engell, James. "Climate Disruption Involves All Disciplines: Who Becomes a Mentor?" In *Teaching Climate Change in the Humanities.* Stephen Siperstein, Shane Hall and Stephanie LeMenage (eds.), London; New York, Routledge, Taylor & Francis Group, 2017. 24–30.

Fielding, Michael. "Students as Radical Agents of Change." *Journal of Educational Change*, vol. 2, no. 2, 2001, pp. 123–141.

Grady-Benson, Jessica, and Brinda Sarathy. "Fossil Fuel Divestment in US Higher Education: Student-Led Organising for Climate Justice." *Local Environment*, vol. 21, no. 6, 2016, pp. 661–681.

Hanson, Nicholas, and Alexa Elser. "Equity and the Academy: A Survey of Theatre Productions at Canadian Post-Secondary Institutions." *Canadian Theatre Review*, vol. 165, no. 165, 2016, pp. 35–39.

Kozelj, Josh. *"Hundreds of Victoria Students Join Protestors in Global Climate Strike."* *The Martlet*. 27 September 2019. https://www.martlet.ca/hundreds-of-victoria-students-join-20-000-protesters-in-global-climate-strike/ (accessed August 27 2020).

Lenferna, Alex. "Divest–Invest: A Moral Case for Fossil Fuel Divestment." In *Climate Justice: Integrating Economics and Philosophy*. Oxford University Press, 22 November 2018. *Oxford Scholarship Online*. https://www-oxfordscholarship.com. myaccess.library.utoronto.ca/view/10.1093/oso/9780198813248.001.0001/oso-9780198813248-chapter-8 (accessed 20 June 2020).

Lewis, Helen. *"When Will We Want to Be in a Room Full of Strangers Again?"* *The Atlantic*. 12 May 2020. https://www.theatlantic.com/international/archive/2020/05/theater-survive-coronavirus-art-west-end-broadway/611338/. (accessed 27 August 2020).

Mansfield, Katherine C. "How Listening to Student Voices Informs and Strengthens Social Justice Research and Practice." *Educational Administration Quarterly*, vol. 50, no. 3, 2014, pp. 392–430.

Mason, Lisa R. *People and Climate Change: Vulnerability, Adaptation, and Social Justice*. New York, NY, Oxford University Press, 2019.

McConachie, Bruce. "Ethics, Evolution, Ecology and Performance." In *Readings in Performance and Ecology*, edited by Wendy Arons and Teresa J. May. New York, Palgrave MacMillan, 2012, pp. 91–100.

McKibben, Bill. *"What the Warming World Needs Now is Art, Sweet Art."* *Grist*. 22 April 2005. https://grist.org/article/mckibben-imagine/ (Accessed 27 August 2020).

Renaud, Lissa Tyler. "Training Artists or Consumers? Commentary on American Actor Training." In *The Politics of American Actor Training*, edited by Ellen Margolis and Lissa T. Renaud. New York, Routledge, 2010, pp. 76–93.

Sanson, Ann V., Judith Van Hoorn, and Susie E. L. Burke. "Responding to the Impacts of the Climate Crisis on Children and Youth." *Child Development Perspectives*, vol. 13, no. 4, 2019, pp. 201–207.

Season Archive. *"Phoenix Theatre Website."* 2020. https://finearts.uvic.ca/theatre/mainstage/archive/ (accessed 26 June 2020).

Siperstein, Stephen, Shane Hall, and Stephanie LeMenager. "Introduction." In *Teaching Climate Change in the Humanities*. Stephen Siperstein, Shane Hall and Stephanie LeMenage (eds.), London; New York, Routledge, Taylor & Francis Group, 2017, pp. 1–13.

Solga, Kim. "Introduction: Theatre and Performance, Crisis and Survival." In *Theatre and Performance in the Neoliberal Academy: Responses to Academy in Crisis*, edited by Kim Solga. New York, Routledge, 2020, pp. 1–8.

"SH." Anonymized Student Interview. 12 June 2020.

"EW." Anonymized Student Interview. 11 June 2020.

"EB." Anonymized Student Interview. 11 June 2020.

"LF." Anonymized Student Interview. 11 June 2020.

"LA." Anonymized Student Interview. 11 June 2020.

"KD." Anonymized Student Interview. 11 June 2020.

"MP." Anonymized Student Interview. 12 June 2020.

"TN." Anonymized Student Interview. 12 June 2020.

Taylor, Diana. *Performance*. Durham and London, Duke University Press, 2016.

Thunberg, Greta. *No One is Too Small to Make a Difference*. New York: Allen Lane, 2019.

UVic. *"Theatre at UVic."* *YouTube*. 4 May 2018. https://www.youtube.com/watch?v=o-GyoxZ8towc (accessed 26 June 2020).

"UVic Ranked as a Global Leader in Climate Action." 2020. https://www.uvic.ca/news/topics/2020+impact-rankings+news. (accessed 26 June 2020).

"UVic Student Society Endorses Student Walkout for Climate." 25 September 2019. University of Victoria Student Society Press Releases. https://uvss.ca/media/news/ (accessed 18 August 2020).

Worthen, W. B. "Acting, Singing, Dancing, and so Forth: Theatre (Research) in the University." *Theatre Survey*, vol. 45, no. 2, 2004, pp. 263–269.

Part 2

Playwriting and collective storytelling

5 Conrad Alexandrowicz and David Fancy in conversation with Caridad Svich

Conrad Alexandrowicz and David Fancy

We decided to interview noted playwright Caridad Svich because of her engagement in environmental and social justice issues, and because, as a teacher of creative writing, she crosses the divide between professional practice and pedagogy. We began our Zoom conversation on July 3, 2020 by noting that while many artists, including those in theatre, are engaging with the climate crisis in their work, very few in the theatre academy seem to be following suit—at least thus far.

DAVID FANCY Caridad, you speak about how explorations of wanderlust and dispossession and notions of migration, both physical and spiritual, are dominant impulses for your plays. Could you talk about how these perspectives inform the work on playwriting and the climate crisis that you've been undertaking for some time?

CARIDAD SVICH It's amusing to me that when you've been writing for a long time you start to see patterns in your work, and reasons why you've circled around specific topics and concerns. I grew up mostly in various eastern seaports—with New Jersey, Florida, North Carolina all being pretty formative—and then eventually moved to California. A lot of the trips that my parents and I made when we kept moving were trips by car, and there's this thing about seeing the United States from the highway. You see it from the backyards, from the edges—not the front. Public facing, but not the shiny side. The sort of the places that feel both abandoned or disused, but used in a specific way in terms of advertising. For example, my own obsession with billboards comes from being on the passenger seat and clocking how many billboards I was seeing. They were all trying to sell me something. I kept wondering about that when I was a kid and I sort of didn't really put it together until later, when I started to think about consumerism and capitalism. So that being said, my first "official" play was called *Waterfall* and it's set next to a toxic landfill. What I wanted to do was to write something that had to do with the overall dysfunction in society and to use that landfill as a metaphor for larger dysfunctions. But I was also thinking about how this affected the people living next to the landfill and how their health was being affected. Why this was built next to them? Why was their neighbourhood considered the place to dump things? As I've been writing, more and more of my work has sort of centred on a connection to or disconnection from the natural world. And who is extracting from the natural world and profiting from it, and who isn't, and why? And, of course, issues of environmental racism sort of rise to the fore through all of this. I love dealing with the natural elements of my writing and putting the audience in spaces where they're connected to the natural elements, making site-specific work or site-responsive work that kind of opens that up in terms of conversation with the audience and their actual immediate environment in terms of where the show is happening, but also exploring ways that the subject itself is extensively around climate grief and climate change.

CONRAD ALEXANDROWICZ You've actually given me a segue into another question. How do you deal with the grief of young people today who are working in theatre? This is something I'm writing about myself, in one of the chapters in this book, and elsewhere. Specifically, how you would assist them to deal with the emotional effects of the climate crisis in their devised or creative work; their grief, anger, anxiety?

CARIDAD When I work with students on the writing table or at the acting table, or other tables, devising and so forth—I hate being prescriptive—but I think that you can't make work ignoring the climate crisis that we are in. I just think that that's a foolhardy enterprise. I think understanding that that's going to be the backdrop of whatever you do changes how you explore whatever subject matters, whatever characters that you're dealing with—if you're making character-based work—needs to be understood in the context of the climate crisis. I mean, look at the moment we're in. But you know, with COVID today, with everything moving online, I have a lot of theatre students who were graduating and kind of facing the world and saying, now what do I do? And there is no industry. There's nothing. We have severe economic inequality. We have this in terms of theatre specifically. All these effects together make for an exponential effect. And so: how could we think about the future now while we're making work with the possibility that there might not *be* a future? And I mean the future in the sense of thinking about a human-centred future of course.... I always talk to my students about imagining a sort of de-centred human: making work that is de-centred around the human may be the way forward; talking to the students and working with them in terms of imagining what kind of stories they want to make and why, and what kind of conversations they want to have with their audiences. I encourage them to think about the long view on climate change, and also that it's been mainly, greatly, a white-centred ravaging of the planet. So that has to be dealt with, right?

DAVID This echoes with how you've talked recently about the notion of theatre as a kind of commons. I'm thinking of Hardt and Negri's work where they're discussing moving beyond distinctions between public and private spaces, as that is a division established by bourgeois law, and instead thinking of the broader notion of the commons that pre-exists this distinction. Are there particular ways to bring about the notion of the commons in how we educate theatre students? And perhaps have the theatre itself be, as you're suggesting, a commons more globally, that engages not just human social spaces, but other-than-human realities affected by the climate crisis?

CARIDAD The notion of the commons has always appealed to me. The fact that so many of the spaces that resemble actual commons are privatized complicates this notion. How we can really claim common spaces, hold on to them, remake them, is central to a lot of what I think about when I'm making work, but also when I'm in a classroom situation with students. I think that the reclaiming of the commons and staking—not "staking," I don't think that word is right—but I think making a space where everyone's invited. The other thing that's been on my mind around common spaces is how to think about other kinds of accessibility. For example: what used to be called street theatre. I was talking to an installation artist the other day about like, can't we just make like… I don't know, puppet theatre in the storefront? Things that feel like people can have access to them, and that the makers don't need to rely necessarily on a building or an institution to make them work; to have healing things. I think that we're making things to heal the ills of the world. Then I wonder: where are the spaces where this kind of work can occur; where are the spaces that are public or semi-public where we can interact with audiences in ways that feel vital? Even if it's secretive and even if it's not being publicized and so forth.

CONRAD "Making things to heal the ills of the world" is so powerful and beautiful! And it's echoed by what is said in the other three-way conversation that's part of this book, actually; in the area of applied theatre. One of the themes emerging in this book is that a lot of us are going to learn how to teach differently, teach different *things*, and teach things *differently*. So, I suppose this is about taking guidance from students who want to be able to choose. They're going to feel compelled to become authors of work that's on this topic, right?

CARIDAD Right, well… I think that first of all, it's about doing the research right. And actually walking outside of the theatre building; like talking to people in the sciences who have been doing this work for a long time. Sometimes you just stay within your own tribe and circle and you don't move outside of it. But that's step one; have those conversations during your research. If you need somebody to be a consultant on a project, have them in the room or however they can participate. I think the other key thing is, like I said earlier, to acknowledge that climate change is the backdrop of everything we live. I always use this analogy when I talk with students where I say: you know, you can't read Jane Austen without thinking about the industrial revolution. So, I think the same when you make anything now where the climate crisis is the backdrop. The question then becomes that if the student, the maker, pretends that the crisis is not there, then you start having that conversation with them. It's like, well, why are you not acknowledging this? Why is this not the backdrop? I always go back to the question of the connection to the natural world: does your work connect to or does it disconnect from the natural world? If this is character-based work then are they living in a region where there's cultural erosion? Are there issues of resiliency and sustainability that are factors? How can the piece that you're making actually be sustainable? So, I love to ask my students to dream big and to write impossible things; then somewhere in the back of their minds on a practical level to be thinking about how is the work going to be achieved on stage, or in original theatre, or whatever your means are. Of course, now we're in a situation historically where climate awareness is contagious and all of this is certainly rising in our collective consciousness. And that's where I think that theatre comes from. The earth is a creative breathing apparatus. The theatre is a creative breathing apparatus established between you and the audience. You're also a creative breathing apparatus: a narrative or non-narrative event that is in connection with an audience, and connected with the ground you're walking on. I remember a colleague of mine who did a piece, and for the first day asked the audience to take off their shoes. I felt it was sort of simple gesture, a beautiful gesture and a kind of very vulnerable one. As soon as you're vulnerable to the earth you start to think about how you are walking upon it. And what those footprints mean. So that's something that I think about often in terms of working in the classroom situation.

DAVID That is rich. In many ways it's about extending the awareness that comes up when people declare their positionality around gender, sexual orientation, ethnicity and so forth. You're addressing a much broader sense of positionality that comes into play when we're thinking of ourselves in the context of the climate crisis, in the context of the earth. Could talk about the role that Climate Change Theatre Action might play in postsecondary education; how that type of project could intersect with theatre training programmes?

CARIDAD Yes, and I'd been doing several theatre actions when Chantal, Elaine (Svich is referring to playwright, director and translator Chantal Bilodeau, who has been instrumental in organizing various theatrical projects concerning the climate crisis; and to playwright, screenwriter and dramaturg Elaine Avila) and others and I got together to do the first one, and it kind of mushroomed from there. I began building from what I had done before: I had done a piece around the aftermath of the Deep Water Horizon disaster in partnership with the Earth Institute and the Water-Keeper Alliance. I had also done a project called *Upon the Fragile Shore* around human rights and environmental rights. The notion of environmental rights is something that I wanted to have in the CCTA, and wanted to think about us having it at the top of the conversation in terms of how we make theatre. "Environmental rights" is a phrase that's not often used when we make theatre, so it began from there when we started to get to work on CCTA. And we commissioned far and wide to get folks to write short pieces around the number one existential crisis of our time. And through the dissemination of those pieces at the college level and at the high school level, we sought to have these short pieces function in the room to stir up conversation; to stir up other kinds of actions; to forge and awaken a kind of thinking around this topic. It was initially conceived as a rapid response project to the climate crisis, but not "rapid" in the sense that the climate crisis is very much an ongoing crisis. Ironically, I think the idea of the "rapid response" had to do with the idea that the clock is ticking in terms of things we can do to actually respond to the crisis. It felt like a kind of call to arms, or call to action.

When we bring the pieces into the classroom setting, I encourage students to make their own pieces around these issues and start them thinking about how they can make that kind of responsive work. It has to do with, what is the conversation we're having? Where can the conversation that we're having with our audiences go? Sometimes that takes the form of wanting to give the audience a lecture. And sometimes that conversation is about what is the next action you can take when you walk out of the door of the theatre? Who is the lawmaker that you want to write to? I'm embracing the idea that there are different theories of efficacy around how you use these pieces in the classroom, the theatre, from me to you and back, but always to move outside the room. In my wildest dreams, you wouldn't need to have these climate action theatre pieces, and everybody would just be out and about dealing with the crisis.

DAVID You're thinking very concretely in terms of how to engage advocacy efforts within the overlapping political constituencies that the climate crisis is playing out in: be they municipal, provincial/state-wide, federal and so on. Where are the levers that you can push? I'm thinking of Boal's "Legislative Theatre" as a kind of example: a certain way of operationalizing insights that are generated theatrically, and through theatrical processes, but then are communicated outwards in ways that are intelligible to policymakers and lawmakers.

CONRAD Yes, and the artist and teacher serve as instigators in so many of these processes. And, by the way, Boal's work is addressed by a number of our contributors, no surprise there. So, with that in mind, how has the climate crisis affected your work as an instructor in playwriting?

CARIDAD It's such a mysterious process, because sometimes you're just chasing your own demons. And sometimes you are thinking, why is there a need to write this now? So that's always a big question: I ask, whose story is truly at the centre? And I think that for me, several things are involved in pursuing these questions. Sometimes it has to do with subject matter, literally the content of the work. What does the content look like? What is the shape of it? Because when we're making theatre, we're making shapes. So it's made me think a lot about two things that are related. I don't use them simultaneously, but I think about them differently in terms of making work. One approach is what I called "waterturgy," which is a dramaturgy that's based on water. The movements of water around ebbs and tides, around currents—structuring plays that sort of mimic water patterns, ocean patterns. In other words, thinking less about land-based and territorial ways of constructing work. And then the other is the notion of "planturgy." In the last two years I've just been really obsessed with "planturgies": plays that branch out; that actually mimic the life of plants; and that reproduce plants in that way, in terms of the storytelling. Think of it this way: you have to change the mode of storytelling in order for people to really see the work. For example, if you're doing land grab mode in your playwriting, an audience is going to stay in land grab mode. I think that you have to make or change the structural mode of storytelling because at the core when you're making theatre it is the structural work that you're giving the audience, and the core has to be changed. Sometimes people will say they're writing a play about climate change, but then they're using modes that are land grab modes, and I'm like, nope: those two things don't go together. You have to think differently about how to structure the story at the root. A lot of playwriting, and especially Western populist playwriting, is territorial writing in which materials and emotions get extracted from other humans and the land that they're in. To change this we need to do metaphorical work, spiritual work and structural work around playwriting.

DAVID By talking about "planturgy" and "waterturgy," there's already a way in which you are introducing a sense that various types of other-than-human entities structure the co-emergence of the work in a kind of intimacy of contact and exchange and relationality between the human artist and the other artists, be they plants or water. You've talked about "planturgy" and "waterturgy" as ways in which humans can be more responsive to the earth in their storytelling. Could you talk about the ways in which other-than-human entities might have the agency, the capacity to give and receive the artistic between one another? How do you include this in thinking about playwriting so that the water and the plants aren't simply a metaphorical resource that are drawn upon, but themselves have some authorship?

CARIDAD Yes, that's something I'm wrestling with right now. I don't have brilliant solutions, partly because I think that because I make theatre, I have a foot in metaphor, a deep sort of metaphor. My other foot as a writer is very interested in installation art. I think where I've staged the most successful work in this area has been in installation art, to be honest. The kind of work where you're asked to visit and listen to the earth for two hours and then read a poem. It's mostly in the live art practices where I've been seeing the most effective work around the climate crisis and the response to it. And I think less so perhaps in playwriting, or more traditional forms of play, where the emphasis tends to be around character and action. And also, our industry tends to prioritize that way of thinking and also the gatekeeping, and the industry prioritizes that as well. What happens to artists that are trying to not work that way is that often they have to make their own company, or they have to leave the industry in order to make the work they want. So, I don't have brilliant answers, but I'm wanting to. I think the most effective for me, when I've done this, have been two specific pieces: there was one about water where we deconstructed a play and actually did it right at a fountain in the town square so that the actors were drenched and the audience was given a visceral experience of water. The other was a piece at Rowan College in New Jersey, where I asked if we could figure out a space on campus where we could do water-based work outdoors. And they suggested a beautiful area of campus near the river; on the bank of the river. During the process the students found out they actually couldn't have performed on that river bank because it was contaminated, but this contamination had been kept a secret. So, sort of this strange thing that happened, where the students were like, Oh, we're on this land and the total area is contaminated: Why isn't the sign posted anywhere? So we decided to figure out a way to do the piece so that people on campus become aware of the contamination.

CONRAD That was such a beautiful, inspiring thing when you talked about "waterturgy" and "planturgy," and the fact that you can't use the same tools to make a different entity. It reminds me of Timothy Morton's idea of "the ecological thought," also the title of one of his books; that we have to transform the way that we think as well as that we have to remediate physical environments that are contaminated. Like that bank you were just describing. With that in mind, what do you have to say about the direction of political and/or environmental theatre in general?

CARIDAD You know, I feel like we're not necessarily at baby stages around this topic, but I feel like sometimes we *are*: what worries me the most is that eco-drama becomes a niche in the academy. And that it's kind of like something that's over there and has no relationship to anything else at work anyway. I see that sort of happening already. It's like, are those people doing drama? As in: some people do real theatre, and then there's this other thing. But really, it's like all drama is eco-drama; all drama deals with the environment, whether the humans are involved in it or not, you know what I mean? My worry is that it becomes sort of a niche thing in the field; not just in the academy, but in the field in general. I'm trying to figure out ways, especially when I'm talking with students, to not see climate-responsive work that way, and to think about how we can embrace... like it's all at the same table in a way that feels equitable. This ties in with how I feel right now, and how I'm so interested in coexistence. A lot of my work has to do with how do we coexist: how do as humans coexist with nature, and how is nature not a static element but actually an evolving and mutating element that is volatile? Because that's the one thing around eco-drama and eco-theatre that sometimes gets misread. The weak idea that nature is kind of placid. It's like, Ooh let's listen to the star. When really there's deep fire happening within the star and it's volatile and mercurial and… full of wrath. And that we are alive, we are organisms, making theatre with other organisms, if that makes sense.

DAVID Absolutely. And this is lot of work around assemblage thinking, where you have an open-ended processual, complex relationality; an open-ended process. That's a lovely articulation.

CARIDAD And it's interesting because I think audiences sometimes get really upset by this; like actually really affronted. I've just seen audiences be affronted by the notion that you can't have a human individual at the centre of the story.

CONRAD Absolutely! And so, with all of this in mind, and all the challenges, both with the climate crisis and also the resistance to dealing with it in our theatre culture: do you have any hope?

CARIDAD I wouldn't write if I didn't have hope. I think writing is a hopeful act. Because you think that it will matter, or you think that it will have an audience; you think that it will last; maybe that's what writers always think about. Did it have a life somewhere? I think that that's central. So, yes, I have hope. Do I have hope about where we are in terms of the climate crisis? I have less hope around that because I just don't see much action occurring at the policy level. In fact, I think we've gone back; obviously this current American administration has gone backwards. I also think there are actionable steps that can be taken. But I don't think that recycling is the only one, and I think that is what the conversation is reduced to. I think that one of the jobs of art is to walk an audience through something. We are walking through an experience together. So, if you're dealing with the trauma, the mostly human-made trauma of climate change, and you're walking the audience through that and making them feel all sorts of things—maybe guilt, maybe fear, or doubt—maybe I want to put my head in the bucket and never go outside again. Whatever they're feeling, how do you allow an audience to move through that experience, and out of that experience into the world? So maybe it's not a matter resolution, which I think can be false, but instead, Well, here are three things that you could do today. I like to leave audiences with questions at the end, or a charge, or a kind of call to action. I like giving them something that they can sort of carry with them beyond the art piece itself. I think maybe that's a way to navigate this question of hope without a bit of a false or easy hope, which I resist a great deal.

DAVID Thank you for that, because it's a nice way of negotiating the spaces between hope on one hand, but not falling into an unjustified optimism. And so, there's a separation of hope: it doesn't necessarily have to be tied into a predictable optimism of a happy ending. But this work of affirmation, of continuing to create, continuing to work through experience, living, creating, teaching the problem, you know, is lovely, lovely stuff. Thank you so much. Is there anything left on that we shouldn't leave on the table that we should talk about?

CARIDAD Maybe the notion of choral. And this is tied to the idea of a collective, like an ensemble protagonist or, de-centring the individual at the heart of a story, if you're telling story. For example, I've just been really, really interested in the idea of a choral mode actually being the way forward, including having the audience be part of that chorus... like handing the audience text to read. And that they're part of that with the actor, with the performers. And if we're making a theatre piece, I'm interested in how we can extend that notion of the chorus, and not think of it as entirely antiquated.... I mean, it's an old idea, but I think of re-reawakening that idea of the choral, and that we're all kind of voicing and articulating in a space. And I like having the audience take, like... not give them roles, but give them texts and allowing them to be at the table. And not as a kind of gimmicky, Hey, y'all get to participate, kind of thing, because I hate that. I think that's kind of silly, but more in a holistic idea that we're all in the same space together when we're making a piece, whether it be virtual or physical or wherever we are. That there's an opportunity for that at some point; I think this important.

DAVID I am thinking of the homonym of choral, spelled C-O-R-A-L with these kinds of collective communities, "planturgies" or "waturgies" It has an interesting resonance for thinking about ways in which we can take existing traditions and reframe them within the context of these other-than-human collectivities... that we don't want to reduce ourselves to necessarily, but they do provide a different sort of inflection or alienation point about thinking differently.

CARIDAD Yes: communities within communities. Communities within and beyond communities. A kind of dynamic theatre where everything is included, everything has its place.

6   Devising in the era of climate crisis

Staging the "eco-performative"

Conrad Alexandrowicz

The topic with which I am concerned here, instruction in the making of climate activist theatre—what I propose to call the "eco-performative"—goes to the heart of fundamental philosophical issues of both ethics and aesthetics that pertain not solely to theatre, but to all the arts; it bristles with questions about basic meanings and purposes and summons a huge body of texts in diverse but overlapping disciplines. Where to begin?

In this intervention in the climate crisis I believe we must consider the promptings—both current and anticipated—of our student cohorts in terms of curriculum planning, as Alexandra Kovacs explores in her contribution to this volume. This is yet another example of the unprecedented conditions thrust upon us by a global emergency: ordinarily we as faculty members are responsible for pedagogical design. But our sense of service to the generation most affected by climate change, while being least responsible for it, as Greta Thunberg is fond of reminding world leaders, compels us to question this precedence. Therefore, as pedagogues working in the areas of devising—what was gathered under the term *collective creation* in earlier decades[1]—we must support the likely desire of theatre students to engage as activist artists with climate struggles. We must be prepared to impart techniques by means of which they may create work that embraces all the possibilities of "political engagement," in various genres and styles, and including everything from work that grapples with explicitly ideological subject matter, to overtly activist work, to acts of protest and disruption that may, strictly speaking, be "illegal." Many young climate strikers, inspired by Thunberg, will be driven to action regardless of the consequences, because, as she observes, "[t]his is above all an emergency, and not just any emergency. This is the biggest crisis humanity has ever faced" (87).

I must make some observations regarding that which is legal or otherwise in this struggle, while noting that it is remarkable that such a discussion, a matter pertaining to legal philosophy, is set down in a volume about theatre pedagogy; it is a measure of its unique purpose. Instructors must be aware that distinguishing between what is "illegal" in law from what is "necessary" *despite* such law, and therefore in accordance with some directive *other than the law*, may be a matter of perspective, as surprising, or even shocking, as this may seem. First, our legal system is largely constructed to protect all kinds of property, including, of course, that of major extractive corporations whose enterprises may be seen as "criminal" from the point of view of the needs of life on the planet, including human life. Second, Indigenous peoples may have a very different concept of what is "legal" from those states that enforce the rights of the aforementioned corporations, and yet Indigenous systems of law may be recognized by those same states that have subscribed to the United Nations Declaration of the Rights of Indigenous Peoples (UNDRIP), including Canada.[2] Therefore our students, both now and in the future, may be more concerned with how to design and mount an effective action, than whether it is, strictly speaking, *legal*, and regardless of whether they may be subject to arrest as a result. As Thunberg has urged, "it is now time for civil disobedience. It is time to rebel" (11).

As I write this, in the midst of a pandemic, and as the effects of the climate crisis continue to accelerate, I remind the reader that theatre is pre-eminently a social art form that depicts relationships, and that therefore treats of ethical questions. These, as Nicholas Ridout has written, "cannot be separated from the specific historical circumstances in which they take place" (7). Further, this function is squared in the conditions of performance: the meanings of ethical questions are amplified in the co-presence of performers and spectators: "We watch ourselves watching people engaging with an ethical problem while knowing that we are being watched in our watching" (15), by both performers and fellow spectators. The deprivations necessitated in the battle against the pandemic have reminded us of all that we take for granted regarding the normal functions of theatre, which are not available at present to represent elements of either of these linked crises.

The "eco-performative"

Theresa J. May, one of the co-editors of *Readings in Performance and Ecology*, coined the term "eco-dramaturgy," denoting "theater and performance making that puts ecological reciprocity and community at the center of its theatrical and thematic intent" (Arons and May 4). In this chapter, focusing more purely on activism, I consider the question of devising the *eco-performative*. This summons a cluster of questions concerning the ethics and aesthetics of politically motivated theatre, including its instrumentalization as activist intervention. I trace some of the history of such theatre-as-activism in the West, including Soviet Agitprop, Brecht's Lehrstücke, Canadian Workers' theatre, Boal's (1979) Theatre of the Oppressed and British protest theatre, in an attempt to situate climate activist theatre as part of a lineage: what can we extract from historical precedents that might be useful to us in our responses to this crisis, one that leftist theatre makers from the last century would not have been able to imagine? I then note some resources that are available to instructors new to supervising student devising, with the understanding that teachers will seek out such resources and use and adapt them according to their needs and preferences.

Using the term "eco-performative" requires that I remind the reader of the precise meaning of "performativity" in contradistinction to "performance." Linguist J.L. Austin developed the idea of *performative* language, that is synonymous with the action it signifies, such as the "I do" of a marriage ceremony—the most widely cited instance (Lecture II, 12–13). In *Utopia in Performance: Finding Hope at the Theater,* Jill Dolan expanded this function of language to include the action of performance itself; that is, she chose to consider performance *performative,* conflating the terms with specific intention rather than using them interchangeably. She used the term "utopian performative" to denote moments that, "in their doings, make palpable an affective vision of how the world might be better" (6). However, as Hans-Thies Lehmann (2006) argued in his influential *Postdramatic Theatre,* while the theatre is able to reveal the truths behind authoritarian political discourse (177), representing "politically oppressed people ... on stage does not make theatre political" (178). Indeed, representing oppression may figure in a larger project of obscuring, neutralizing or excusing it: "It is not through the direct thematization of the political that theatre becomes political but through the implicit substance and critical value of its mode of representation" (ibid.). We may anticipate that student concern will likely not reside in making "political theatre" but rather with "making theatre political" (Ridout 65), in the ways both Dolan and Lehmann theorize. "Eco-performative" captures the *functionality* of performance events, in which I include the disruptive actions of Extinction Rebellion that both exhort and effect a transformed attitude towards the whole of the global biota, thereby subverting the nature/culture and human/nonhuman divides.

If activism is the sole remaining morally responsible function for theatre as an art form—as our engaged students of the future may argue— then "climate theatre" will tend to be realized in terms of the "eco-performative"; but will that disqualify it as art? Has not "art" always been distinguished from such activist interventions as agitprop—considered further below—that may be dismissed as nothing more than "propaganda"? Artists, as I know from much experience, frame questions in the process of what may be described as "uncovering" their work, without necessarily engaging in the conscious process of doing so. They configure complex and ambiguous images and arrive at finished objects that are susceptible to multiple and even contradictory interpretations. Propagandists proceed with the answer(s) firmly and unquestioningly in hand and seek only to convert audiences to their beliefs, and thereby to compel action. Their "art" is fully instrumentalized and goal-oriented. If "poetry" may stand for "art," then the last line of Archibald MacLeish's "Ars Poetica" is noteworthy here: "A poem should not mean, but be." Accordingly, the more "poetic" a work of art, including a work of theatre, the more its value lies simply in its *being,* and the less it may be said to align with "propaganda," which is largely a vehicle for formulaic *meaning* intended to produce particular effects.

Editor Nina Felshin's collection of essays entitled *But Is It Art? The Spirit of Art as Activism* dates from 1995 and considers this topic in the context of visual, conceptual and performance art produced in the second half of the 20th century. But the features and conditions of creation of these expressions align with our understanding of such work in our own art form: the work tends to be process- rather than product-oriented, to be physically situated or site-specific, to be collaborative in nature and to be integrated within networks and communities (10–12). It tends to exploit the possibilities for media exposure and coverage (16)—in our time this would perhaps be superseded by coverage in *social* media—and to be participatory in structure and function. This last point is charged in terms of the frequently observed disjunction between aesthetics and ideology: such participation may be "impossible if ambiguity and obscurity, however provocative aesthetically and intellectually, bar comprehension" (25).

As for the question of the aesthetic value of politically motivated art, her contributors seem to have arrived at a consistent answer to the ironic question that forms the title: "But does it matter?" (13). Not if one focuses solely on the objective of activism—in this case *ecoactivism*—which is to change hearts and minds, and to spur specific kinds of behaviour. But Sarah Ann Standing (2012), contributing to *Readings in Performance and Ecology*, considers that understanding such action as art is crucial for at least two reasons: first, it places it within a longstanding tradition that includes "situationist art, agit-prop, farce, and performance art, as well as Dada and Futurism"; and second, it can open up possibilities for expression among artists who might not otherwise be inclined towards activism, if it can only be conceived as within their purview, which is "art." (148)

"Art" vs. "propaganda"

Is the art/propaganda distinction in some way a false one? Is not all art prompted by and constructed around some essential message that may be equated to the intention of its author to incite some kind of alteration in the reader/spectator? In *Theatre of the Oppressed* Augusto Boal considers Aristotelian, Hegelian and Marxist notions of theatre's social functions and finds commonality among them: "According to Aristotle, as well as Hegel or Marx, art, in any of its modes, genres or styles, always constitutes a sensorial way of transmitting certain kinds of knowledge" (53).

One thinks, as one compelling example, of the entire centuries-long corpus of Medieval church architecture, stained glass and sculpture, as well as much Early Modern painting, which was intended to instil in an illiterate population belief in the universal truths of Christian doctrine and to condition behaviour accordingly: acceptance of the given orders and conditions of life as the price of nothing less than eternal salvation. As Boal wrote of feudal art, "Its function was authoritarian, coercive, inculcating in the people a solemn attitude of religious respect for the status quo" (55).

A 2019 show by Chinese artist/activist Ai Wei Wei in Düsseldorf was entitled, "Everything is art. Everything is politics."[3] One might argue with the truth of this double claim, in particular its first component, and it is perhaps more convincing, and certainly more pertinent to my discussion, to compress the two and state that "All art is politics," in the sense that it either endorses the status quo or questions it. Lehmann would certainly contend that all theatre is always already a political gesture:

> Theatre itself would hardly have come about without the hybrid act that an individual broke free from the collective, into the unknown, aspiring to an unthinkable possibility; it would hardly have happened without the courage to transgress ... all borders of the collective.
>
> (179)

Entertaining this radical notion may prompt us to revisit the origins of theatre in our own cultural inheritance. What was the "point" of the great plays of the Athenian dramatic efflorescence? (Those few that have come down to us.) According to Boal, Aristotelian tragedy is essentially coercive: its core function, that of catharsis via the emotions of pity or fear, amounts to "the purgation of all antisocial elements," a function, he argues, that "survives to this day, thanks to its great efficacy" (46). Theatre historian Jennifer Wise suggests that many of these plays were essentially courtroom dramas and addressed problems of justice such as those argued in the city's proliferating law courts; that both the law and theatre arose together as discursive practices that were dependent upon alphabetical writing (119–168). And while Greek drama and comedy dealt with a whole slew of questions concerning how we ought to live, there was one overriding and unifying directive in the plays, one that has particular bearing on my subject: "Because the Greek gods were nature deities, all of the divine moral imperatives to "respect the gods" that we hear in every tragedy (and some comedies) are really, almost literally, ecological imperatives: respect nature. That's the inviolable law of the gods." (Email communication)

Considering audience reception

Before examining some historical examples, I consider some tools for ana-lysing theatre in general, including the subset of "activist" theatre. Jerzy Grotowski asked, "Can theatre exist without an audience? At least one spectator is needed to make it a performance" (32). How that spectator responds to the theatrical event summons the notion of the "horizon of expectations," employed, among others, by aesthetic philosopher Hans Robert Jauss, an early contributor to the field of "reception theory."[4] The following definition, while not directly from Jauss, is concise and useful:

> The shared 'mental set' or framework within which those of a par-ticular generation in a culture understand, interpret, and evaluate a text or an artwork. This includes representational knowledge of con-ventions and expectations (e.g. regarding genre and style), and social

knowledge (e.g. of moral codes). It is a concept of reading (and the meanings this produces) as historically variable.

(Chandler and Munday)

How audiences respond to climate-activist performance will depend on the set of assumptions they bring with them to the performance venue: the degree to which they share the ideological position of its makers, their notions regarding how such a work ought to be configured, including the degree to which they believe that theatre/performance ought or is equipped to address the topic, and questions regarding their expected response to it.

Baz Kershaw's (1992) *The Politics of Performance: Radical Theatre as Cultural Intervention* is concerned with four decades of activist theatre in England, from the 1960s to the '90s, and is both a work of theoretical reflection and of local theatre history. Citing the work of sociologist Elizabeth Burns, Kershaw notes that every work of theatre entails an "ideological transaction" between performers and audience (16). He stresses the importance of context, "the propensity of a performance text to achieve different meanings according to the context in which it occurs." (ibid.)

Susan Bennett has also focused on the crucial constitutive role played by theatre audiences, in particular in non-mainstream theatre, in which we can certainly locate the eco-performative. As she observes, the dramatic text itself is subject to change according to audience response, in rehearsal, previews or even during a run (20). As will be demonstrated, activist theatre tends to be based in collage and multimedia and, therefore, has even more moving parts that may be subject to change after being tested by audiences.

Some historical instances

We have numerous models from which to draw inspiration in imagining how devising might be focused on climate action, and, more broadly, on a theatre of environmental activism (given that climate change is only the direst of a host of interrelated processes of environmental degradation). But it must be said that while there is a vast body of scholarly reflection on theatre *practice*, there is relatively little on theatre *pedagogy*, therefore what we gather from these accounts must be more or less repurposed for pedagogical application.

While all art may be "political"—insofar as it is about the operations of power—not all political art may be considered "propaganda," perhaps the best example of which is agitprop. As Alan Filewod writes, the term referred to expressions in a variety of media and

> originated from the early Soviet conjunction of propaganda (raising awareness of an issue) and agitation (exciting an emotional response to the issue) ... [A]gitprop developed in Russia and Germany as a mobile form of exhortative revolutionary theater designed for quick

outdoor performance. It was adaptive to location, audience, and cast, and suited the sightlines and acoustics of outdoor performance in found spaces.

Agitprop arose in a time of revolutionary ferment, before the arrival of electronic media, when theatre could still function as an effective means of communication, when class divisions were relatively clear and straightforward and when armed struggle against the bourgeoisie was the unquestioned objective.

Kershaw notes that in the second half of the 20th century in the West, "conditions of cultural pluralism produced by political consensus, relative affluence, and the ameliorating force of the mass media" (80), agitprop as a form became problematic. But theatre-as-political-poster-art has a reinvigorated potential in the third decade of the 21st century, when the contradictions—to use Marxist-Leninist terminology—of the climate crisis have become acute and all-encompassing. They amount to a battle between the forces of life and death, as Greta Thunberg is constantly reminding the world's business and political elites. Agitprop asks no questions, but merely presents a set of *a priori* assumptions that is assumed by all, actors as well as audiences; it demands consensus and urges action. This means, as Baz Kershaw wrote, citing David Edgar, that agitprop removes one of the most powerful tools of the dramatist, what the Greeks called *peripeteia*, or "a change of attitude in a character, or in the audience's attitude to a character" (Kershaw 79). It also forbids irony, which entails ambiguity and therefore the potential for query (81). These provisions perhaps sum up quite neatly what distinguishes art from propaganda.

Arising after Soviet agitprop theatre were Bertolt Brecht's *Lehrstücke*, literally, "teaching pieces." This term, as Erika Hughes writes, "describes a series of experimental works written in the 1920s and early 1930s by Bertolt Brecht and a number of collaborators, including Kurt Weill, Hanns Eisler and Elisabeth Hauptmann" (197). Citing Frederic Jameson, she notes that these experiments, with their open-ended construction, susceptible to audience intervention, seem to herald the theory of Paulo Freire and its manifestation in the forum theatre of Augusto Boal. Noted Brecht specialist Willett concurs, noting that the Lehrstücke were animated by "the notion that moral and political lessons could best be taught by participation in an actual performance" (33). On the matter of aesthetic pleasure vs. moral instruction, Brecht maintained that "theatre remains theatre even when it is didactic, and as long as it is good theatre it is also entertaining" (80). And, in keeping with much 20th century practice, Brecht noted that the learning plays summoned interdisciplinary composition, including the use of film and music. The overall aim was to "show the world as it changes (and also how it may be changed)" (79).

The practices in Augusto Boal's *Theatre of the Oppressed*, considered by a number of contributors to this collection, led to and became known as "Forum Theatre." They belong in the realm of applied theatre, as they are deployed in particular communities and rely on their members' participation as both audience and performers—as "spectactors" (*Games* 254). It is however, instructive to note how his methods align with those we employ in teaching performance students how to make their own work, and how such work might be activist in disposition, such as his recommendation that all such practices begin with the body. (126)[5]

Marxism-Leninism was, of course, an international movement, that acquired considerable adherence and traction during the Great Depression. The global economic disaster—the like of which we are seeing as a result of the COVID-19 pandemic—prompted workers' theatre movements in Canada and the US, as well as in Europe. Such companies often performed outdoors and on picket lines. As David Gardner observes, the most renowned "production of the Workers' Theatre was *Eight Men Speak* (1933), a full-length play based on the trial and imprisonment of eight Canadian communists," which was banned in Toronto and Winnipeg.

Carnival vs. agitprop

While overtly political theatre may be said to emerge as an effect of the growth, spread and political consolidations of Marxism-Leninism, a theatre of resistance, a "people's theatre," is much older and is linked to the history of carnival. Theorized in literary studies as "the carnivalesque" by the influential Russian critic Mikhail Bakhtin, it is concerned with celebration, excess, sensuality and the body, and with the inversion of prescribed orders. In *Hyperion and the Hobbyhorse* Arthur Lindley writes:

> Carnival, for Bakhtin, is an embodiment of the liberated communality of the people in perennially renewed rebellion against the social and spiritual restrictions of the official order ... [it] also celebrates the dominance of the feminine ... [and] is the healthy assertion of the rights of the body, the material principle, at the expense of the spirit.
> (17, 18)

This contrasts with the cerebral, instructive disposition of ideological theatre that seeks to contain and direct an audience's attention and that may be gendered as masculine. An effective critique, meant to be received by an audience in a state of sober reflection, may be thoroughly dissolved in the joyous and sensually gratifying—and *participatory*—celebrations of the carnivalesque.

While noting that agitprop and carnival were the "chalk and cheese" of British alternative theatre, Kershaw also observes that "they represent the ends of a single spectrum ... and which unites them despite their differences" (68). In his case studies of later work in this period he finds that perhaps the most effective political theatre is that which is located in the middle of this spectrum and is thus able to combine celebration—including comedy and satire—with pointed critique (220–242). Coining the term "carnival agit-prop," he describes a genre where celebration was combined with protest "in different degrees according to the needs of a particular context for which a performance was designed" (82). Students devising eco-performative theatre need to grapple with the opposition of these two terminal points, while considering the opportunities that may arise given that they are indeed part of a spectrum. Where will they choose to situate their work?

Kershaw documents the work of playwright Ann Jellicoe, whose efforts in the 1980s "almost amounted to a new genre: the community play" (175), which in retrospect we recognize as another precursor of applied theatre: Such performance work was instigated by professional animateurs, including the playwrights that Jellicoe hired, but was created with, by and for particular citizen constituencies to address socio-political issues specific to their experiences. Her work, according to Kershaw, had to tread very carefully as its large scale necessitated broad community involvement across divisions of class and sex, among others. In the aim to create unity her "plays aim to transcend social differences, at least temporarily, and to use participation in creative work to strengthen the networks of community" (190). Its results were therefore more celebratory than subversive; material differences in communities that are revealed by the explicit representations of agitprop tend to be concealed by the festivities of carnival, in which everyone is invited to participate. Jellicoe's critics felt she was glossing over real differences of status and power in the effort to produce such celebrations, in which a temporary illusion of equality was created. (ibid.) This subject area is the terrain of those contributors to this collection who work in applied theatre, where the issue of acquiring and exercising sensitivity to the realities of differences within communities is of crucial importance in the pedagogical process. How will students create meaningful climate theatre in communities where the issue is still highly contested and there is real potential for conflict?

One of the plays under Kershaw's consideration is *The Cheviot, the Stag and the Black, Black Oil*, written by John McGrath in collaboration with members of the 7:84 Theatre Company in 1973. The name was derived from a statistic in *The Economist* in 1966: 7% of the world's population was said to own 84% of its wealth. (McGrath 76) (Note that this ratio is even more skewed at the commencement of the third decade of the 21st century.) The piece was a "ceilidh play," combining scenes, storytelling and a great deal of music, and documented the clearances that took place in Scotland in the 19th century, when thousands of people were thrown off the lands they rented from aristocratic landlords to make way for a breed of sheep called the Cheviot. Over the course of three tours to locations all over Scotland, the work was performed for approximately 30,000 people and was made into a film for the BBC (McGrath vi, xxvii). The work connected the history of the clearances with the discovery of oil in the North Sea, in which a new story of dispossession began to unfold, and foreign interests were essentially allowed to steal the lion's share of the profits while exploiting the local work force. These lines from the show connect the two struggles: "We too must organise, and fight—not with stones, but politically, with the help of the working class in the towns, for a government that will control the oil development for the benefit of everybody" (73).

There are a number of instructive elements in the account of this play, not least of which is the ironic reversal of the value of fossil fuels in a work of protest theatre from almost fifty years ago, before the catastrophic effects arising from their use had become clear. But the play offers many lessons for those who seek to create similarly engaged work, in particular its use of music, which has tremendous power to draw and engage an audience. Theatre students with musical ability ought to consider how these skills may serve the goals of activist intervention. Further, its subject matter grew from and was rooted in an issue of compelling local interest, was developed from meticulous research and conveyed with respect and compassion for the audience whose history it sought to illuminate. And finally, each performance concluded with a dance for the spectators and continued for as long as attendance warranted, even if the company had an early departure the next day. Providing the audience with a fully embodied, explicitly celebratory experience—after they had spent the better part of the evening seated, as audiences usually are—raised the temperature of community support for the whole endeavour.

Another of Kershaw's provocative and instructive cases is that provided by Welfare State International, the remarkable performance group renowned for producing "carnivalesque agit prop" on a massive scale (212–242). In 1979, after spending many years in urban settings, as well as touring to international festivals, the company relocated to a small town on the south coast of Cumbria, and in 1983—when the US was installing cruise missiles in Europe, increasing Cold War tensions to an almost unbearable degree—the group commenced a seven-year residence in the nearby town of Barrow-in-Furness (213). One of the town's principal industries at the time was the construction of Britain's Trident nuclear submarines. As Kershaw asks, "how was a company with such radical objectives going to be received in such a town?" (214). Intriguingly, their first work was a full-length feature film for approximately fifty unemployed local young people, commissioned from playwright Adrian Mitchell and based on Shakespeare's *King Lear*. Entitled *The Tragedy of King Real*, and featuring "a kind of punk-medieval rabble in a wasted industrial junkyard" that suggested a post-holocaust wasteland, "the film had a blatantly anti-nuclear story" (215). The work relied on complex intertextuality, presuming the audience's ability to engage in sophisticated simultaneous reading, and made reference to fairy tales and Christmas pantomime, among other texts (216); as in the case of *The Cheviot...* music and songs were also crucial components of the work (217). Kershaw notes "that *King Real...* met with a fairly cool public response in Barrow," but credit was given for the creative outlet it gave a large number of local youth, and, in its wisdom, the borough council "saw the promise of more widespread cultural animation" (220). That is, the virtues of the event itself served to mitigate its less than enthusiastic reception.

Some years later, in response to the relative easing of Cold War tensions, the company produced *What'll the Lads Do on Monday?* Written by Albert Hunt and devised with a local cast of eighteen, the title was widely recognized as a quip by a local Conservative MP: "If Labour gets in on Friday, what'll the lads do on Monday"? (230) The play employed the formal methods of collage and montage and included many songs, lyrics from one of which ran as follows:

What'll the lads do on Monday
Now that the Cold War has gone
What'll the lads do on Monday
And all other days from now on
There'll be no more pay in the packets
The orders are all down the drain
So what'll the lads do on Monday
Now that the world has gone sane (Hunt, qtd. in Kershaw 234)

All one need do is replace "Cold War" with "oil" to see how both the methods and materials of a work from forty years ago may be applicable now, as we face the biggest environmental crisis in our species' short tenure on this earth.

The large-scale performance works that Welfare State created with residents of the town, and in various of its public spaces, offer valuable examples of how trenchant satire may be combined with democratizing celebration so that critique may be productively deployed. Both ethical and aesthetic values may facilitate the communication of highly subversive and provocative ideological messages. Such work forms "a subtle example of theatre as a weapon" (Kershaw 242).

The remarkable success of Welfare State's work in Barrow may be attributed in large part to the conditions that a long-term residency made possible. Perhaps a group of student actors—or a company of recent graduates—might engage in a summer residency, creating and presenting a play about the politics of fossil fuel production in communities along the route of the much-contested Trans Mountain pipeline expansion in British Columbia. How might a work of climate-activist theatre—addressing the issue of climate justice as well as the catastrophic effects of continued fossil-fuel burning—find a productive reception in such communities? What balance might be struck, as Welfare State was able to do in Barrow, between celebration of community, visual spectacle and pointed critique, without the audience either walking out or obstructing the performance? As Kershaw urges, "to stop the unthinkable we have to think the unthinkable" (219), whether the matter is challenging English shipbuilders to consider that the nuclear weapons they are producing could wipe out all life on earth, or Canadian workers, laying pipe for the dirtiest oil in the world, to question whether such fuels ought to be left in the ground because burning them will lead to planetary death via a different route. (Note that no one expects that acts of theatre will convince anyone to quit one kind of job and seek another!)

On the question of efficacy

The question remains: even if one insists that they ought ethically to do so, can performance events actually *produce* change? Do we act *differently* in response to what we have experienced as members of an audience? One of the keywords of Kershaw's book is "efficacy": To what extent can "the micro-level of individual shows and the macro-level of the socio-political order ... somehow productively interact" (1). Kershaw makes a series of recommendations that climate theatre makers would do well to heed: The more localized and functionally embedded theatrical interventions are the more they might prompt actual change (246). And theatre artists will be more likely to succeed if they are oriented away from the "conventions of mainstream theatre ... [in order] to make performance more accessible" (246) and to make reference to other theatre forms, as well as non-theatrical social events. As important as context is intertextuality. Perhaps the most successful of the works Kershaw analyses use intertextual references, in particular to well-known works, and in doing so achieve two seemingly contradictory goals: to create complicity between performers and audience, but also to create room for "forms of discourse which can be disruptive and dangerous" (247).

Do the aesthetic values of eco-activist theatre amplify its potential performative effects? A number of contributors to *Readings in Performance and Ecology* support this proposition. Sarah Ann Standing, writing about the remarkable installation by Earth First! called "Crack the Damn," claims that "[t]he power of ecoactivism as art is grounded in its aesthetics. Whether or not it moves people to change personal behaviors or public policies depends fundamentally on its artistry" (154). Anne Justine D'Zmura, documenting *Green Piece*, a work of eco-dance/theatre/installation made with students at California State University, Long Beach, recalls how "[a]udience members were visibly moved by their experience and ... spoke of desires to become proactive in their community to help facilitate change" (178) Bruce McConachie, writing about ethics and ecology, and citing the foundational work of John Dewey, claims that "imaginative engagement in the arts provides real experiences that change who we are and can motivate progressive change in the world" (98) And, as Boal famously wrote: "Perhaps the theatre is not revolutionary in itself; but have no doubts, it is a rehearsal for revolution!" (155).

However, much late twentieth-century theoretical speculation on theatre's capacities in this regard is weighted in the opposite direction. Hans-Thies Lehmann reminds us that theatre operates "in a world of media which massively shapes all perception" (185) and by means of which representation is divorced from real experience: we switch from one channel depicting the misery of Rohingya refugees in Bangladesh to another featuring a reality show about competitive weight loss. Lehmann argues that the most theatre can do in the face of this is to engender in the spectator "an *aesthetic of responsibility (or response-ability)*" (ibid., emphases in original). Nicholas Ridout interprets such *response-ability* in terms of both perception and the capacity for action: in the co-presence that is performance

> [s]pectators are called upon to recognise that there is a relationship between what is shown in the theatre and their own experience of the world. In responding to this call, spectators take responsibility for making what is shown part of their personal experience.
>
> (59)

I suggest that, given the evidence of praxis over the course of the twentieth century, it is more likely that activist theatre is a *consequence* of broader social and political action rather than its instigating force. And perhaps its greatest value is its capacity to generate a sense of unity and purpose in particular communities, as John McGrath noted in retrospective consideration of the remarkable phenomenon that was *The Cheviot, The Stag and the Black, Black Oil,*

> [t]he theatre can never *cause* a social change. It can articulate the pressures towards one, help people to celebrate their strengths and maybe build their self-confidence. ... Above all, it can be the way people can find their voice, their solidarity and their collective determination. If we achieved any one of those, it was enough.
>
> (xxvii, emphasis in original)

Climate guardians

The aforementioned objections about the need for art to survive its reduction to propaganda, as well as questions concerning its efficacy, may be rendered moot in the climate crisis. As I ask above, does the sole remaining respectable—or even *possible*—function for theatre pedagogy lie in training actors to create and perform events such as those staged by the Extinction Rebellion[6] protesters? Performative by definition, they generate as much resistance to as support for the cause they represent.

To *witness*, to *protest*, to *disrupt*: these are distinct but overlapping verbs, and they perhaps form a logical sequence. Witnessing can become a form of protest, albeit a passive one; protesting, need not, but *can* progress to outright disruption. Extinction Rebellion protestors generally disrupt everyday activity—"business as usual"—in order to enforce awareness of the climate crisis on their captive audiences. But, as Denise Varney has written, the "Australian Climate Guardians present a novel approach to political performance ... with the unexpected turn to religious iconography" (67). Participants from a variety of backgrounds, mostly women, but some men as well, dress as angels, complete with feathered wings and long white robes, and undertake silent acts of witnessing to protest inaction in the face of the crisis, "up to 60 or 70 at any one action" (ibid.). Most famously, they assembled at the Eiffel Tower during COP 21, which resulted in the Paris Accord. (ibid.) The angel is a compelling invention of Judaism, Christianity and Islam; a messenger from a higher power, often a warrior, both human and superhuman, embodying some ultimate moral directive, and standing guard over something precious, in this case the earth itself. In the actions of the Climate Guardians aesthetic and ethical values are fused, and such spectacles perhaps have more efficacy as acts of protest theatre than the purely functional disruptions staged by Extinction Rebellion. Here, once again, I note that theatre students will have to consider such differences of disposition in their eco-performative interventions.

Would such acts of civil disobedience—and only these—satisfy militant student climate strikers who choose post-secondary education in the-atre as part of their operational toolkit? We dismiss art that qualifies as "propaganda" because it presents only one point of view, simplistically, one-dimensionally; we consider that it diminishes the potentials of the art form and insults our intelligence. But in the case of an overwhelming and unprecedented crisis, presenting differing points of view on matters of fundamental rights, including the right to existence itself, establishes a false equivalency that is ethically untenable. By means of this logic we do not give a platform to white supremacists or Holocaust deniers. The complaint of false equivalency can be made regarding Israel/Palestine: it is misleading to speak of "the two sides in a conflict" when one party not only has all the military, economic and juridical power, and therefore controls the destiny of the other, but also controls the means of production and dis-tribution of the historical narrative of the subject, and therefore its mean-ing. In this programme of erasure, as Edward Said wrote, "Palestine does not exist, except as a memory or, more importantly, as an idea, a political and human experience, and an act of sustained popular will" (5). In some predicaments there *are* no shades of grey, and the question then becomes: which of two very clearly opposed sides will one support? As Thunberg said at the Davos summit in January 2019,

> [y]ou say that nothing in life is black or white. But that is a lie. A very dangerous lie. Either we prevent a 1.5°C of warming or we don't. Either we avoid setting off that irreversible chain reaction beyond human control—or we don't. Either we choose to go on as a civilization or we don't. That is as black and white as it gets. There are no grey areas when it comes to survival.

(19–20)

Practically speaking...

Many books have been written on devising in the past few decades, a num-ber of which I note for the reader's guidance, assuming she or he will search out those that seem of greatest interest and utility. But first it must be repeated, as is noted in the Introduction, that this practice is not new. Arising in the 1970s, as part of the broad cultural revolution that began in the preceding decade, and practised by a number of companies, it was called "collective creation," and resulted in a remarkable body of work that remains instructive today, in particular in terms of the conception of the actor as both creator and interpreter.

- *Devising Theatre: A Practical and Theoretical Handbook*, by Alison Oddey, dates from 1994 and is the first practical book of its kind.
- *Devising: A Handbook for Drama and Theatre Students*, by Gill Lamden, published in 2000,
- *The Viewpoints Book: A Practical Guide to Viewpoints and Composition*, by Ann Bogart and Tina Landau, has become widely used since it was published in 2005. It is as much a program of movement-based improvisation as it is a guide for collective creation.
- *The Frantic Assembly Book of Devising Theatre* by Scott Graham and Steven Hoggett is both a historical account of work by the celebrated company and a practical guide.

Notes

1 For a historical review of such practices see *Devising Performance: a Critical History*, Deirdre Heddon and Jane Milling, Palgrave Macmillan, 2006.
2 See https://www.aadnc-aandc.gc.ca/eng/1309374407406/1309374458958 (accessed 20 April 2020).
3 https://www.maxhetzler.com/news/2019-05-18-ai-weiweieverything-art-every-thing-politics-solo-show-k20-k21-kunstsammlung-nrw-dusseldorf-18-may-1-september-2019 (accessed 2 April 2020).
4 See https://www.oxfordreference.com/view/10.1093/oi/authority.20110803100407730 (accessed 15 June 2020).
5 Boal's story is indeed remarkable: while he was artistic director the Arena Theater of Sao Paulo became one of South America's leading leftist theatre companies. A military dictatorship came to power in Brazil in 1964 and "[h]is outspoken position against the authoritarian regime led to his imprisonment and torture in 1971" [*Theatre* 156]. While eventually released and acquitted of all charges, he was forced to flee the country with his family and moved to Buenos Aires. He was then forced to leave Argentina, also for political reasons, when the military took power in that country—he may be the only theatre artist in history to be exiled twice for dissidence—and settled in Portugal.
6 See https://www.dw.com/en/british-parliament-declares-climate-change-emergency/a-48568627 (accessed 2 May 2019).

Works Cited

Arons, Wendy and Theresa J. May. "Introduction." In *Readings in Performance and Ecology*, edited by W. Arons and T.J. May, Palgrave Macmillan, 2012.

Austin, J.L. *How to Do Things with Words*. Clarendon, 1962.

Bennett, Susan. *Theatre Audiences: A Theory of Production and Reception*. Routledge, 1990.

Boal, Augusto. *Theatre of the Oppressed*. Theatre Communication Group, 1979.

———. *Games for Actors and Non-Actors*. 2nd ed., translated by Adrian Jackson, Routledge, 2002.

Brecht, Bertolt. *Brecht on Theatre: The Development of an Aesthetic,* translated and edited by John Willett, Hill and Wang, 1964.

Chandler, Daniel, and Rod Munday. *A Dictionary of Media and Communication.* 2nd ed. Oxford UP, 2016. (accessed 19 February 2019).

Dolan, Jill. *Utopia in Performance: Finding Hope at the Theater.* U of Michigan Press, 2005.

D'Zmura, Anne Justine. "Devising Green Piece: A Holistic Pedagogy for Artists and Educators." In *Readings in Performance and Ecology,* edited by Wendy Arons and Theresa J. May, Palgrave, 2012.

Felshin, Nina. "Introduction." In *But Is It Art? The Spirit of Art as Activism,* edited by Nina Felsin, Bay Press, 1995.

Filewod, Alan. "Agitprop." *Routledge Encyclopedia of Modernism.* 2017. https://www.rem.routledge.com/articles/agitprop-theatre (accessed 23 July 2019).

Gardner, David. "English Language Theatre." *The Canadian Encyclopedia.* 2006. https://www.thecanadianencyclopedia.ca/en/article/english-language-theatre (accessed 8 May 2020).

Grotowski, Jerzy. *Towards a Poor Theatre,* edited by Eugenio Barba, Methuen, 1980.

Hughes, Erika. "Brecht's Lehrstücke and Drama Education." In *Key Concepts in Theatre/Drama Education,* edited by S. Schonmann, SensePublishers, 2011.

Kershaw, Baz. *The Politics of Performance: Radical Theatre as Cultural Intervention.* Routledge, 1992.

Lehmann, Hans-Thies. *Postdramatic Theatre,* translated by Karen Jürs-Munby, Routledge, 2006.

Lindley, Arthur. *Hyperion and the Hobbyhorse: Studies in Carnivalesque Subversion.* University of Delaware Press, 1996.

MacLeish, Archibald. "Ars Poetica." *Poetry Foundation.* 2020. https://www.poetryfoundation.org/poetrymagazine/poems/17168/ars-poetica (accessed 7 April 2020).

McConachie, Bruce. "Ethics, Evolution, Ecology, and Performance." In *Readings in Performance and Ecology,* edited by Wendy Arons and Theresa J. May, Palgrave Macmillan, 2012.

McGrath, John. *The Cheviot, the Stag and the Black, Black Oil.* Methuen, 1981.

Ridout, Nicholas. *Theatre and Ethics.* Palgrave Macmillan, 2009.

Said, Edward. *The Question of Palestine.* Vintage, 1992.

Standing, Sarah Ann. "Earth First!'s 'Crack the Dam' and the Aesthetics of Ecoactivist Performance." In *Readings in Performance and Ecology,* edited by Wendy Arons and Theresa J. May, Palgrave Macmillan, 2012.

Thunberg, Greta. *No One Is Too Small to Make a Difference.* Penguin, 2019.

Varney, Denise. "Hosts of Angels: Climate Guardians and Quiet Activism." In *The Routledge Companion to Theatre and Politics,* edited by Peter Eckersall and Helena Grehan, Routledge, 2019, pp. 66–71.

Willett, John. *Brecht on Theatre: The Development of an Aesthetic,* translated and edited by John Willett, Hill and Wang, 1964.

Wise, Jennifer. *Dionysus Writes: The Invention of Theatre in Ancient Greece.* Cornell UP, 1998.

7 Anthropogenic anxiety and the pedagogy of climate crisis in *Wake Up Everyone*

Gloria Akayi Asoloko and Soji Cole

Many countries in the world are focusing significant attention on climate change and its observable alterations in the constitution of the natural environmental design of the earth. This attention has prompted the need for global educational supports characterized by creative programmes in order to ensure sustainable practices and the well-being of communities. Most of these programmes have been sustained by pedagogical approaches which are aimed at promoting thinking and action to effect environmental justice. The Cornell University Climate Smart Solutions Program in the United States, and the Sustainable Schools in New South Wales, Australia, are typical examples of these climate change programmes.

In Africa, there has been consistent fragmentation in the climate change debate. For instance, in Nigeria, the government has over the years developed many policies which concern environmental protection and preservation. However, these developments have been characterized by multiple implementation missteps, lack of sustainable action, cultural inhibitions ingrained in religious and traditional belief systems, and the complete denial of the climate change phenomenon by the majority of the population. In 2010 the federal government, through the Ministry of Environment, drafted a policy document on climate change entitled *National Environmental, Economic and Development Study (NEEDS) for Climate Change in Nigeria*. Right from its introductory pages there is the expression of sentiments which can be interpreted as lack of political will to implement the agenda of the document. On the Executive Summary page, the document states that

> [a]lthough Nigeria has yet to fully undertake detailed assessment of cost estimates for national adaptation actions and programs, available information from various sources in the country indicate that it will be very costly for Nigeria to adapt well to anticipatory climate changes.
>
> (6)

The documented policy itself was only adopted by the government in 2012 under the name *Nigeria Climate Change Policy Response and Strategy*. Consequently, most of the government policies have failed to achieve any comprehensive success. This failure has resulted in trails of unfettered human actions, such as deforestation (especially through bush burning), pollution (institutional gas flaring and oil spills), and general environmental degradation all around the country. Since the government has shown little leadership on climate change, the people of Nigeria have in consequence—generally speaking—not taken responsibility for the crisis as citizens. The only action taken by the government to excite the interest of the Nigerian people on climate change has been ceremonial tree-planting exercises. Unfortunately, due to its immense fossil fuel extraction, Nigeria continues to contribute significantly to anthropogenic climate change in Africa.

Theatre has provided one of the significant modes which individuals and agencies have employed to generate communicative and discursive platforms for the climate change conversation in Nigeria. A study of these modes reflects the implementation of Theatre for Development—a generally accepted term in Africa for "applied theatre"— initiatives as a mainstay of the theatre modalities used to engage with climate change in communities in Nigeria. This seems to have some positive value, especially in comparison with the way purely scientific data is communicated in the media, in which there is little or no opportunity for feedback. To effect positive and sustainable action on climate change, the Theatre for Development initiative in Nigeria has also been limited by its own inherent attitude: over-reliance on the community to propagate and take action. The workflow in Theatre for Development usually takes the steps of organizing and planning, meeting with the community, meeting/selecting volunteer performers, casting, script development, rehearsal and performance. Since Theatre for Development follows certain structured processes which align with group engagements, it is usually difficult to engage it in a pedagogical approach in which individual responses and follow-up actions are as important as the group's objectives. Written drama, as a theatrical mode, can function in the intrapersonal and interpersonal dimensions. It can be utilized as a pedagogical tool to enhance individual initiatives, or produced as performance to promote community education on climate change.

In this chapter we explore the critical pedagogical potential of *Wake Up Everyone*, a drama by Nigerian scholar Greg Mbajiorgu, which, in its consciously designed structure, is an instrument for teaching and learning. It constructs and mobilizes environmental imageries that are transformed into privileged modes of understanding within the cultural milieu of Nigeria, thus closing the knowledge gap on the subject of climate disaster. The drama explores the anthropogenic origins of climate change that are supported with scientific data; *Wake Up Everyone* thus dramatically emphasizes the connection between the people and their environment for Nigerian audiences. Its enables understanding of the four key points in the narrative of climate change described by Al Gore and referenced by Dale Anderson:

1. The overall average temperature of Earth's atmosphere has been rising rapidly in recent decades.

2. Human activity—mainly the burning of fossil fuels, which produce greenhouse gases—is the chief cause of that temperature rise.
3. If temperatures continue to go up, humans will see a host of natural disasters, including droughts that will ruin crops, severe storms that will damage property and kill people, and sea levels that will rise to flood low-lying coastal areas and islands.
4. People can prevent these disasters, but they must start taking steps to do so immediately (6).

Wake Up Everyone is thus considered in this chapter as a valuable source of pedagogical intervention; as a kind of "knowledge pamphlet" that is crucially important in a society with limited conversation and action on the hazards of linked environmental crises. We adopt the methodology of content analysis of the primary text, which is subjected to interpretation in the context of key interdisciplinary texts on subjects related to climate change.

Nigeria's climate has undergone drastic and inconsistent changes in the past few years. While this is not unexpected, because of large-scale human activities which have placed the country as one of the top greenhouse gas emitters in Africa, the surprise is the lack of response to this unfolding disaster. From 2009 to 2014, the Government of Canada partnered with the government of Nigeria, and substantially funded an environmental project named "Building Nigeria's Response to Climate Change." The project was implemented by Cuso International and International Coaching Federation, and the Nigerian Environmental Study/Action Team (NEST). On the Government of Canada's website, the project is described as seeking

> to enhance Nigeria's ability to reduce poverty in an equitable and sustainable way by putting in place more effective governance related to climate change. The main activities of the project involved conducting research studies on the best strategies to adapt to climate change; undertaking community-level adaptation pilot projects; supporting the development of a Nigeria Climate Change Adaptation Strategy; developing and using climate change education and outreach materials; and developing gender-specific tools and mechanism.
>
> (Gov't of Canada: Nigeria)

While there is currently no available information to clarify if a post-project implementation review was undertaken, in order to track response to the work by the communities involved, it is clear that the efforts exerted as part of the project, especially in the area of educational outreach on climate change, did not continue. This claim can be supported by the persistent ignorance of the subject by the majority of Nigeria's populace, and the lack of any policy framework on climate change, while the country has been experiencing increasingly devastating and widespread effects of the climate crisis since 2015. Air and water pollution, vector-borne diseases, flooding and other damage to agricultural lands have been some of the resulting consequences of the climate crisis in Nigeria. Flooding is the most prevalent of these disasters, massively ravaging whole sections of the country every year. In the foreword to a technical report entitled *Nigeria Post-Disaster Needs Assessment 2012 Floods* by the Federal Government of Nigeria, it was disclosed that

> [f]loods are the most common and recurring disaster in Nigeria. The frequency, severity, and spread of these floods are increasing. Beginning in July 2012, heavy rains struck the entire country. The impact of the 2012 flooding was very high in terms of human, material, and production loss, with 363 people killed, 5,851 injured, 3,891,314 affected, and 3,871,53 displaced.
>
> (iii)

Both in terms of the effects on health and the loss of produce and land, the farming community in Nigeria is usually the sector of the country worst hit by the climate crisis. Idowu et al. reported the following effects in terms of the public health of Nigeria's farming communities, who make up over 70% of the population:

- Respiratory diseases due to increases in the level of pollutants
- Malaria (in more widespread levels within the population 70% annually)
- Skin ailments (45% annually)
- Heat stroke (4% annually)
- Loss of productivity (40% annually)
- Potable water shortages (60% annually) due to floods and/or saltwater intrusion. (148–149)

One of the reasons why there is massive unpreparedness on the part of these farming communities is lack of knowledge about the climate crisis. This ignorance has been reinforced by religious and other traditional belief systems, which have consistently been in conflict with notion of the values of formal education. Extreme religious beliefs and practices of Christianity, Islam and Indigenous traditional religion have blurred the boundary between spirituality and scientific communication. The narrative construction of the climate crisis in Nigeria represents the power of these belief systems to dominate and direct the course of social subjects. One of the dominant flaws of these belief systems is the inability of their believers to create room for ontological distinctions between humans and the events which surround them. Such systems usually push social issues into a discursive framework within the rhetoric of "mob psychology," a situation where the mental energy of people is unquestionably directed only towards the cause with which their belief system aligns. The consequence of this is that conventional pedagogical strategy is bound to fail in the objective of making crucial connections in the context of climate change education available to local communities. As a consequence of its privileged position in the cultural milieu, drama can be utilized as a useful pedagogical method of dismantling configured cultural belief systems and the religious steadfastness which have undermined the majority of Nigerian publics from acquiring a "ventilated" knowledge of the dynamics of the global climate crisis.

The debate on the climate change crisis has shifted considerably, from outright denial to acceptance of its anthropogenic origins, and more recently, to what can be done to respond to and mitigate the phenomenon. However, climate change may be difficult to conceptualize, because it is not directly experienced, except in the form of its consequences, weather events themselves, which, however extreme, can usually be included within the normative. Chaudhuri and Enelow, whose work is cited elsewhere in this volume, have stated that "[t]he only way it can be apprehended is through data and modelling—through systems and mediations—all of which have to be processed cognitively and intellectually: have to, in short, be understood, rather than experienced, phenomenologically and temporally" (23). As Chaudhuri, Enelow and many others have argued, this notion of phenomenal conception be articulated through the pedagogical methodology of drama; a method that utilizes the aesthetic values of both text and visual imagery in its explorations. Although certainly influenced by our perspective as theatre practitioners, we feel that while other modes of narrative (film, novel, etc.) have enabled the anxiety and trauma of the climate crisis to gain traction in the popular imagination, that it is more difficult to deploy them as pedagogical tools for the learning and teaching of climate change subjects. We believe that drama, because of its "liveness," can more readily establish itself beyond the purview of imagination to that of interaction, experience and action. Amitav Ghosh's suggestion that "[i]t is as though in the literary imagination climate change were somehow akin to extraterrestrial or interplanetary travel" (7), is a typical recognition of the inability of other genres of fiction to fully express the reality of the climate crisis beyond the imaginary. Drama, as a discipline, and as a mode of pedagogy, possesses the capacity to help learners generate viable hypotheses. Hypothesis, as a potent component of pedagogy, opens up channels of possibilities for trial and error. Enlightenment philosopher John Locke reflected that "hypotheses, if they are well made, are ... great helps to the memory, and often direct us to new discoveries" (642). Drama as a methodology for teaching is capable of bringing the learner's reality into transformative possibility. It is capable of creating a more vivid imagination for learners to independently problematize and process the texture of knowledge. Drama can thus help decolonize knowledge, if deployed in specific ways, and make the teacher more the purveyor of knowledge than its sole "producer." This contrasts with "[a] modern view of empowerment [that] perceives students as being ignorant about their state of freedom, not really aware of reality until the teacher as emancipator makes a 'powerful intervention' to free them from their previous state of ignorance" (Biesta 55). Students utilizing the pedagogical tools of drama tend to have more confidence in developing active voices in the classroom environment. In its narrative, communicative and dialogic methodologies, drama as an educational tool is a meaningful and critical form of pedagogy between learners and teachers, as many of our co-contributors also affirm. Using drama as a form of teaching and learning on the climate crisis therefore empowers drama teachers to become what Henry Giroux has termed "transformative intellectuals" (125). Drama enhances the dialogic quality of relationship that is created both inside and outside of the classroom; this is essential in teaching and learning about the climate crisis.

Though environmental issues are considered as an aspect of education for Nigerian students, they are never an emphasized form of study except in the sciences. Even then, attention is primarily directed towards environmental issues regarding the oil-rich Niger Delta areas of the country. This appears to be the foremost objective, rather than the need to recognize other problems resulting from climate change. Tarila Ebiede, in his article *Conflict Drivers: Environmental Degradation and Corruption in the Niger Delta Region,* claims that the Niger Delta area is critical in financing the continuing pattern of corruption and personal aggrandizement of public officers and their cronies at the expense of needed public services, institutions and infrastructure. Most Nigerian dramas have therefore focused their thematic preoccupation on the environmental issues of the Niger Delta areas. This area of focus only takes into consideration the politics of oil and oil extraction, and the consequent corrupt dealings by government, institutions and powerful individuals. The real dangers of climate crisis portended by oil extraction are seldom tackled in the dramas.

Greg Mbajiorgu's (2011) *Wake Up Everyone* represents a bold step in addressing this neglected area of the Nigerian environmental crisis. The play portrays a situation where "[t]here is mounting evidence that climate change is and will continue to negatively impact health in many countries" (World Bank xii), and that the world is at risk from the consequences of human actions which are producing the climate crisis. The drama underscores the anxiety that results from the looming dangers of the climate crisis, but is as well as a radical call to take action to stem some of its worst effects. Owing to the scanty supply of teaching materials on climate crisis in Nigeria, the play can be regarded as a novel pedagogical experiment. After the drama was published, the author, who is a drama teacher at the University of Nigeria, Nsukka (UNN), was awarded the 2017 Sustainability Award for Environmental Education by the UNN-SHELL Centre for Environmental Management and Control (CEMAC). Immediately afterwards, the vice-chancellor of the university appointed him a member of a university committee entitled Building Trans-disciplinary Climate Change Capacity. These achievements may be attributed to the drama's pedagogical capacity to demonstrate how viable solutions may be put into practice on the subject of climate change; urging the need to chart strategies for learning about the subject and engaging with it in curricula. Perhaps these efforts will form some part of an overall project of public education.

Wake Up Everyone is set in "Ndoli," a fictitious farming community in Nigeria. Prof. Aladinma, a professor of agriculture who doubles as a theatre practitioner, is the central character in the play. He runs a theatre troupe in his village alongside his vocation as an agriculturist and university teacher. The author has deliberately configured this central character as an agricultural scientist in order to situate the drama in a realistic context. The character's disciplinary background leaves no room to discountenance his intelligible commitment to scientific enlightenment. The complex challenge of the climate crisis demands that dramatic interventions involve resource-focused pedagogical methodology so as to enlighten both students and the broader community on the subject. Prof. Aladinma stresses the particular impacts of climate change on agriculture. He sensitizes the local farmers on safer practices for farming in a bid to adapt to the changing climate. His dramatic enactments in the community become his major tool for sensitization and training of the majority population of farmers. As stated above, farming in Nigeria is a vocation that is particularly affected by the climate crisis every year. Statistically, also noted above, those who engage in farming in Nigeria form more than half of the entire population of the country. It is therefore crucial to the play's meaning that its locale is a farming community. This is also important as these communities belong to the constituency of Nigeria's population that has the least outreach in terms of education, and more particularly the lack of awareness of climate change. Prof. Aladinma attempts to influence the government to act in a timely fashion to forestall imminent unprecedented disasters by funding and implementing essential environmental programmes. Unfortunately, the government, as represented by the local chairman, Hon. Edwin Ochonkeya, nonchalantly disregards Prof. Aladinma's warnings. In the end, the village suffers a heavy flood which has disastrous human and material consequences.

The dialogue in *Wake Up Everyone* instigates an appraisal of gaps in knowledge that inspire further thinking. The play presents three categories of character in the framework of climate change awareness. The first might be called "the informed and concerned":

PROF. ALADINMA: You see Mr. Chairman, when the drummer changes beat, the dancer must change his steps. Things are no longer the way they used to be, and even a child must have noticed the changes in our climate. Take the rain for instance, the downpour this year has been heavier than that of any other year. From news reports around the world, natural disasters have become daily occurrences, earthquakes, tsunamis, hurricanes, floods, and many others (13).

Second, "the informed and unconcerned":

CHAIRMAN: (Cuts in) But these are expected, Prof. Aladinma. Doesn't the Bible say that in the last days things like these will happen, and…? (13).

For this second category of people, there is an awareness that the disasters of climate change are likely, but that they have little or nothing to do with anthropogenic activities. The belief is that such occurrences are bound to happen anyway and there is nothing that can be done about it. The third category of people might be described as "the absolute ignorants." For this category of people, any such disaster only happens if a certain divine force is angry with the desires of humans: One of the farmers in the play, **Mazi Chinedum,** states, "[t]he gods have cursed the rivers of Ndoliland" (62). The reluctance of the characters to acknowledge the reality of climate change is grounded in their ignorance of the issue and the absence of some transformative pedagogical structure to counter such ignorance and engender knowledge. **Mazi Chinedum** makes the following assessment:

> Well, after he spoke to us, I did not take him seriously. I mean, the man was sounding like a prophet of doom. He kept talking about how man has made the world to be in its present bad state, and how man will finally destroy it if he continues this rape of the planet.
>
> (67)

Drama serves as a critical tool for orientation in this kind of "knowledge dilemma." Whether as creative praxis, classroom dialogic communication or in its theoretical expression, drama-based pedagogy aims at creating an imaginary world between learners and teachers. "In this respect we can say that knowledge results from the cooperation of experience and action" (Biesta and Burbules 55). In reference to examples in the UK, Manfred Schewe notes that "[o]ver the last two decades, drama pedagogy has helped to lay the foundations for a new teaching and learning culture which accentuates physicality and centers on 'performative' experience" (19). The combination of performative experience with learning culture is important in situations where lived or imagined reality serves fundamental pedagogical functions, rather than being dependent on theoretical speculation.

In Prof. Aladinma's theatre rehearsal in *Wake Up Everyone*, many affected by the crisis experience suffering and anxiety:

NWEKE: Here in Ndoli, our source of living has gone forever, our fishermen have exhausted what is left of our sacred waters, carting home tender fingerlings that hold the secret of future harvest.

OBIOMA: And now that we have completely destroyed the immune system of our planet, there is weeping and wailing everywhere and fear of increasing extremities. From one end… (As she points to one end, Nweke dramatizes human calamity caused by flood.)

NWEKE: Help! Help! I am drowning: I am dying, save me, save me!

OBIOMA: And from the other end… (As she points to the other end, Ekene dramatizes human agony in times of drought.)

EKENE: Oh! Give me water, give me water or I die, I'm drying up, I am thirsty. Ah! I can see the shadow of rain, God send down the rain, send down the dew of life. (49–50)

Such anxiety is widespread, as Prof. Aladinma himself reflects in a conversation with another character:

PROF. ALADINMA: "They forgot to list the recent floods in Sokoto, Gombe, Kano, Katsina, Bayelsa, Maiduguri, Calabar, Lagos, Ibadan, and we may face one here in our very own land if we fail to take the bull by the horns" (58).

The character's recounting of the devastation caused by flooding in major cities in Nigeria represents the author's deliberate manoeuvring of Nigeria's lived reality into the context of the drama. This "artistic insinuation" is justified in a country where it is difficult to influence the government's position on climate change intervention, as well as to effect attitudinal change among the people. (A couple of years after the play was published almost all the states in Nigeria experienced devastating floods.)

The play also points to the fact that the climate crisis is an amplifier of social problems in Nigeria that range from public corruption and poverty to rising inequality. It depicts scant prospect of relief as there has not been evident general public awareness of the climate situation. As the play unfolds a foreign student arrives in the village on a research trip on the subject of climate change. The professor introduces the student to his theatre troupe: **"Prof. Aladinma**: Desmond is a PhD student of the University of Manchester, North-east England. He is here to carry out a field work on Drama and the War Against Climate Change in Africa. Interesting, isn't it?" (53) While the student finds the Professor's theatre education programme pertinent to his study, the latter is upset that attention has not been paid to the same issue of climate change in his own society: **"Prof. Aladinma**: In this country, we have done nothing, absolutely nothing, and our leaders who should show more concern are completely nonchalant" (60).

Wake Up Everyone signals that drama-based educational content can strengthen the architecture of climate change knowledge in Nigeria; as a performance project, it can reasonably be read as part of a classroom syllabus and for training outside the classroom. This play, and others like it, could be incorporated into theatre study in higher education in Nigeria. It aligns with the position taken by Chaudhuri and Enelow: that drama can bring the kind of radical re-imagination that is necessitated by the perilous predicament in which we find ourselves and our fellow species.

Disseminating knowledge on the climate crisis may not be enough; however, as learners must be engaged in ways where they will not only acquire knowledge but also the opportunity to re-create such knowledge. In his treatise on the autonomy of the classroom, Paulo Freire, cited by Maria Vargas, notes that

[t]he very first of these types of knowledge, indispensable from the beginning to the teacher (that is, to the teacher who considers him or herself to be an agent in the production of knowledge), is that to

teach is not to transfer knowledge but to create the possibilities for the production or construction of knowledge.

(11)

There is a fervent need to rouse epistemological curiosity in students, most especially students in post-secondary education, who are the appropriate population that can engage others on knowledge of climate change, and in particular in countries such as Nigeria. James Nehring suggests that educators have ascribed the terms "teacher-centred" and "student-centred" (30) to two distinct strategies in pedagogy. It is worth revisiting the differences in these approaches given that, in terms of climate activism, a "public pedagogy" will need to be deployed in the context of entire populations, derived from the purely academic version.

As is well established in the literature, the teacher-centred variety is described as the traditional model, where the learning situation places the instructor in a position of control over the students, the materials of study and the ways in which students learn. The teacher is the most active persona in this learning scenario. The students are passive learners, listeners and merely receptors of knowledge. It is an intra-active pedagogical structure.

However, student-centred learning typically gives students more agency as learners. It establishes avenues for students to participate more actively in discussions, the designing of learning projects, exploration of topics of interest and sometimes configuring their own course of study. This is an inter-active pedagogical structure of learning. Both methods have merits and limitations, and drama-as-pedagogy incorporates elements of each. It informs learning in ways that promote experience-based perceptions, that awaken feeling and sensory states and that create meanings that link the personal to the cultural. This approach does not tilt the weight of pedagogical activity in one direction or the other. It accords rather with Leo Tolstoy's opinion that

> [t]he school will, perhaps, not be a school as we understand it—with benches, blackboards, a teacher's or professor's platform—it may be a panorama, a theatre, a library, a museum, a conversation; the code of the sciences, the programme, will probably everywhere be different; so also with the pedagogical methods of drama. (Nessel 23)

Much reflection on the pedagogy of drama in the climate crisis begins with the understanding that the humanities—especially drama—can work effectively with the sciences in promoting knowledge of the existential threat posed by the overall environmental crisis. (A topic that Lara Aysal also explores in this volume.) The end result of a product of science, whether in its pure or applied form, will ultimately fail to benefit humanity, if it does not aspire towards having relationship with other disciplines that would directly connect it to the human society. Latour suggests this in his prescriptive essays on modernity when he remarks, paradoxically, that "[t]he more science is absolutely pure, the more it is intimately bound up with the fabric of society" (43). Michel Foucault (2003) also argues in *The Birth of the Clinic: An Archaeology of Medical Perception* that ways of thinking in modern medicine have changed significantly because of the reformulation of the ground of knowledge in other disciplines, such as linguistics, economics and philosophy (14). Both Latour's and Foucault's proposed conceptual frameworks are not simply about the classifications and epistemologies of science and other disciplines; they are also about cultural and political representation in disciplinary interconnection. The productions of drama, in both their written and performed manifestations, possess the capacity to integrate inputs across disciplinary divisions.

The objective of this chapter is not to presume the integration of drama with the sciences in their entire or sub-disciplinary domains. This would certainly be a difficult and time-consuming project, both in processing and procedural formalities, as Pickett et al. have suggested: "The conceptual framework of each subdiscipline must be well formulated, and a new framework that links both must be developed. This requires careful attention to the basic focus, or domain, involved" (14). What is suggested in this chapter is that drama can serve very well as a veritable conveyor in articulating what the sciences have unearthed, or what they predict. Drama can bring to light the results of the work of science in ways that citizens will fully understand. Drama, in this conception, can be regarded as a concise and significant pedagogical strategy for expanding knowledge about climate change.

Practically speaking...

- Plays about climate change should engage the authenticity of social and cultural norms of a particular locale or target audience. As many have argued, including contributors to this collection, a human story is more likely to capture attention and engage the imagination than a rendition of verified scientific data.
- The story should not be unnecessarily burdened by the dissemination of information on climate change. In other words, the aesthetic values of the play should be considered as important as actual "information."
- On the other hand, plays about the climate crisis need to use metaphors to enhance understanding, rather than being deployed for their own sake. Drama is no exception among the genres of writing that thrive on metaphor. But for the sake of efficacy—a topic addressed elsewhere in this collection—climate activist drama should consciously enlist metaphor to enhance understanding, rather than as puzzle that prompts multiple meanings. It will serve no purpose to write the equivalent of *Waiting for Godot* as a drama on climate change if the function of the lone tree in the play (in fact a significant object to consider as metaphor for climate change!) will continue to be the subject of diverse speculations by the audience, even years after the play was written.

Given that the authors of this chapter are also playwrights, we speculate that playwriting students can consider writing plays on the climate crisis following these steps:

- Conduct preliminary research with a view to construct a convincing and captivating narrative. Such research, in the conditions we describe above, might be based on (a) the understanding of anthropogenic climate change; (b) the lack of climate change literacy; (c) climate change denial; (d) navigating anxiety; and (e) community investment and stewardship in environmental issues.
- Every piece of drama localizes an issue, even if the issue is part of a global phenomenon. By fixing on a target audience, a writer can invent a local character who can tell part of a global story about climate change.
- The language should, understandably, suit the target audience. While the story may be global the language used to convey it must be local.

Works Cited

Anderson, Dale. *Voices for Green Choices: Al Gore – A Wake-Up Call to Global Warming.* Crabtree Publishing Company, 2009.

Bakhtin, M.M. *Speech Genre and Other Late Essays*, translated by Vern W. McGee, University of Texas Press, 1986.

Biesta, Gert. "Don't Be Fooled by Ignorant Schoolmasters: On the Role of the Teacher in Emancipatory Education." *Policy Futures in Education*, vol. 15, no. 1, 2017, pp. 52–73. doi:10.1177/1478210316681202. (accessed 26 May 2020).

Biesta, Gert and Nicholas Burbules. *Pragmatism and Educational Research.* Rowman and Littlefield Publishers, 2003.

Chaudhuri, Una and Shonni Enelow. *Research Theatre, Climate Change, and the Ecocide Project: A Casebook.* Palgrave Macmillan, 2015.

Cook, Justin W. *Sustainability, Human Well-Being and the Future of Education*, edited by Justin Cook, Palgrave Macmillan, 2019.

Ebiede, Tarila. "Conflict Drivers: Environmental Degradation and Corruption in the Niger Delta Region." *African Conflict and Peacebuilding Review*, vol. 1, no. 1, 2011, pp. 139–152. https://www.jstor.org/stable/pdf/10.2979/africonfpeacrevi.1.1.139.pdf (accessed 14 June 2020).

Federal Ministry of Environment Abuja, Nigeria (Special Climate Change Unit). "National Environmental, Economic and Development Study (NEEDS) for Climate Change in Nigeria." 2010. www.unfccc.int/files/adaptation/application/pdf/nigerianeeds.pdf (accessed 6 August 2020).

Foucault, Michel. *The Birth of the Clinic: An Archaeology of Medical Perception.* Routledge, 2003.

Freire, Paulo. *Pedagogy of the Oppressed.* 30th anniversary ed, translated by Myra Bergman Ramos, Continuum International Publishing, 2000.

Ghosh, Amitav. *The Great Derangement: Climate Change and the Unthinkable.* University of Chicago Press, 2017.

Giroux, Henry A. *Teachers as Intellectuals: Toward a Critical Pedagogy of Learning.* Bergin and Garvey, 1988.

Gore, Al. *Earth in the Balance: Forging a New Common Purpose.* Taylor & Francis, 1992.

Idowu, A., S.O. Ayoola, A.I. Opele, and N.B. Ikenweiwe. "Impact of Climate Change in Nigeria." *Iranica Journal of Energy & Environment*, vol. 2, no. 2, 2011, pp. 145–152.

Kolbert, Elizabeth. *Field Notes from a Catastrophe: Man, Nature and Climate Change*, Bloomsbury, 2006.

Kottak, Conrad. "The New Ecological Anthropology." *American Anthropologist*, (New Series), vol. 101, no. 1, 1999, pp. 23–35.

Latour, Bruno. *We Have Never been Modern*, translated by Catherine Porter, Harvard University Press, 1993.

Locke, John. *An Essay Concerning Human Understanding*. The Pennsylvania State University Electronic Classics Series. 1999. www.enlightenment.supersaturated. com/johnlocke/BOOKIVChapterXII.html (accessed 30 April 2020).

Mbajiorgu, Greg. *Wake Up Everyone*. Kraft Books Limited, 2011.

Nehring, James. "Progressive vs. Traditional: Reframing an Old Debate," *Education Week*, vol. 25, no. 21, 2006, pp. 29–42.

Nessel, Denise. *Leo Tolstoy, Education and Culture in Awakening Young Minds: Perspectives on Education*. Malor Books, 1997.

"Nigeria Post-Disaster Needs Assessment 2012 Floods." A report by The Federal Government of Nigeria with technical support from the World Bank, EU, UN, and Other Partners, 2013. www.gfdrr.org/en/publication/nigeria-post-disaster-needs-assessment-floods-2012 (accessed 2 May 2020).

Picket, S., J. Kolasa, and C.G. Jones. *Ecological Understanding: The Nature of Theory and the Theory of Nature*. 2nd ed., Elsevier, 2007.

Plumwood, Val. *Environmental Culture: The ecological crisis of reason*. Routledge, 2002.

Sayne, Aaron. "Climate Change Adaptation and Conflict in Nigeria." *A United States Institute of Peace Special Report, no. 274*, 2011. https://www.usip.org/sites/default/ files/Climate_Change_Nigeria.pdf (accessed 1 May 2020).

Schewe, Manfred. "Taking Stock and Looking Ahead: Drama Pedagogy as a Gateway to a Performative Teaching and Learning Culture." *Scenario Journal of Language, Culture and Literature*, vol. 7, no. 1, 2013. research.ucc.ie/scenario/2013/01/ Schewe/02/en (accessed 3 May 2020).

The Government of Canada. "Building Nigeria's Response to Climate Change." *Government of Canada website.* 2011. www.climate-change.canada.ca/finance/ details.aspx?id=460 (accessed 23 April 2020).

United Nations. "United Nations Framework Convention on Climate Change." 1992. www.unfcc.int/files/essential_background/background_publications_ htmlpdf/application/pdf/conveng.pdf (accessed 25 April 2020).

Vargas, Maria. "Opening Spaces for Critical Pedagogy through Drama in Education in the Chilean Classroom." An unpublished thesis submitted to Trinity College Dublin in fulfilment of the requirements for the degree of Doctor in Philosophy in the School of Education, 2019.

Watson, Nigel. "Postmodernism and Lifestyles." In *The Routledge Companion to Postmodernism*, edited by Stuart Sim, Routledge, 2001.

World Bank Group. "Geographic Hotspots for World Bank Action on Climate Change and Health." *Investing in Climate Change and Health Series*, 2017. www.hdl. handle.net/10986/27810 (accessed 23 April 2020).

World Health Organization. "Climate Change and Health." *Factsheet*, 2018. www.who. int/news-room/fact-sheets/detail/climate-change-and-health (accessed 23 April 2020).

8 "Can we talk?" Forum theatre as rehearsal for climate change interventions

Derek Davidson

We have all been here: families around the dinner table or gathered at holidays slump into chilly silence when one bold family member dares to bring up the subject of climate change. A surly uncle begins to shout that no one should bring up politics at mealtime. The youngest roll their eyes and ask to eat dessert in their rooms. At the workplace colleagues shrug, mumble something about how even scientists cannot agree on whether that whole climate thing exists. We have watched media outlets and pundits frantically contort themselves in order to avoid speaking meaningfully about this new "bogeyman of the liberals." We watch as our political leaders, when confronted with questions about the crisis, change the subject, speak derisively about it, sometimes deny it and sometimes refuse to answer altogether. *Climate change blah blah blah, it isn't real; or if it is, there's nothing anyone can do about it. So let's talk of something else, shall we, as the continents burn and the seas begin to boil?*

The subject of climate change holds talismanic power in its ability to warp, intensify or curb conversations; like some medieval phrase from the Kabbalah, or a magical incantation from a B movie: people jabber, shift uneasily, speak in tongues, slink into annoyed silence, occasionally shout and curse. What is it about climate change, arguably the most pressing issue of our time,[1] which causes such paroxysm?

And yet even as more admit that climate change *is* an important issue, statistics often reveal a persistent refusal to discuss it, at least in much of the United States (where I live and teach).[2] One of our country's leading educational institutions has been tracking this disparity: Yale Climate Connection's most recent map, from 2019, shows that 67% of people in this country believe climate change is real; however, according to the same source, only 36% of people talk about issues of climate change "at least occasionally."[3]

One thing with which most would agree is that talking about climate change is the best thing *to* do, as a means of collectively acknowledging its very real presence, and of better preparing for its inevitable consequences. But overcoming social reticence has proven challenging for numerous reasons, not least of which is because issues surrounding climate change have become increasingly politicized in the United States: in many circles merely broaching the topic without irony is to align oneself with one political party and to incur the rabid indignation of members of the other. I do not wish to suggest that the United States is the only country with a significant percentage of the population denying or downplaying climate catastrophe's imminence; I use my own country rather as representative of the conditions currently under examination.[4] And to return to the image with which this essay began: many students and faculty at my university confess that when visiting family during holiday gatherings—aware that climate change can potentially cause heated arguments or uncomfortable silences—they avoid bringing the issue up at all.

Another reason often cited is that many non-scientists feel they lack the expertise to speak on the subject at all. They do not belong to the scientific community; their knowledge is spotty, gleaned from occasional public radio coverage or random articles skimmed briskly on social media. They presume their ignorance regarding melting polar ice caps, rising temperatures and sea levels and the like can only be adequately explained—or even discussed—by scientists, authorities who understand the complexities of weather systems, species migrations, ocean science and so forth. Moreover, scientists themselves, many of whom having spent most of their lives working in labs or in the field away from society, have not practised communicating with populations outside their specific scientific community. As much as they would like to explain the evidence pointing incontrovertibly to a dire climate future, they do not know *how*.

However, as I suggest in the following pages, one may find the beginning of a way to bridge these challenges of communication in pedagogical practices that have their origin in the theatre arts. I will outline possible solutions that I have drawn from the work of Augusto Boal, the Brazilian theatre practitioner and theorist who strove to employ theatre in a variety of settings as a tool for empowerment (as have other contributors to this volume.) Since the subject of climate change can in its sheer magnitude overwhelm people, giving them a sense of impotence, paralysis, futility—"There's nothing we can do, so why talk about it?"—then Boal's methods of empowerment can effectively fight against the fatalism that leads to crippling silence. But before proceeding to an extended examination of Boal's (1979) Forum Theatre, I offer a brief explanation for how I, a playwright, actor, director and educator, became involved in issues of climate change at Appalachian State University, a mid-sized university (part of the University of North Carolina system) nestled in the Blue Ridge Mountains in western North Carolina.

The climate stories collaborative

Appalachian State University (ASU), with six undergraduate colleges, boasts a solid commitment to environmental issues. Its College of Fine and Applied Arts (CFAA), where the Department of Theatre and Dance is housed, also has a Department of Sustainability, and a Department of Sustainable Technology and the Built Environment. In the spring of 2017

Laura England, a senior lecturer in the Department of Sustainable Development, approached me about possibly co-facilitating an initiative based on the work of author-activist Jeff Biggers, who in February had given a talk at our university, "An Evening at the Ecopolis: Presenting a Regenerative City." Jeff's presentation inspired England to begin the process of creating a group of passionate, dedicated faculty—starting with a small handful of professors from the College of Fine and Applied Arts—interested in dispelling what she termed the "spiral of silence" surrounding climate change, modelling her initiative after similar work being done at Yale and Iowa State.[5] Professor England and I met with the Dean of our college, who enthusiastically endorsed, and even financially supported, the initiative we called the Climate Stories Collaborative.[6]

That September, England and I co-facilitated an introductory workshop, in which approximately twenty CFAA faculty participated; faculty from such diverse disciplines as Sustainable Development, Art, Theatre and even Military Science. In the first of our two Friday workshops, we outlined the issues we aimed to address and explained our mission: to bring scientists and artists together in order to increase collaboration among our disciplines and to open up opportunities for conversations between artists and scientists.

It is important to remember that I, like many in the artistic community, am committed to environmental issues, cognizant of the ever-smaller window afforded our species to combat the worst effects of climate change and dedicated to the idea that creating art engaging these issues can be a powerful and affecting means of raising awareness. But—as with many of my fellow artists—I am *not* a scientist, nor am I especially knowledgeable regarding the complexities surrounding and caused by our changing climate. I know I lack expertise: often I feel like the tourist who, having memorized a slew of useful phrases, will enter conversations I am ill-equipped to understand.

However, my colleague Laura England believed that my position outside the scientific community could prove useful. Theatre practitioners learn early that good collaboration skills and the ability to communicate among diverse groups are necessities for artistic success. Collaboration and communication: among the challenges our Climate Stories Collaborative is confronting is learning to help scientists come up with better communication strategies to reach a larger public, and for non-scientists to find the freedom to discuss fears engendered by our rapidly changing climate, and to discover, explore, cultivate ways they might respond to those changes. The particular skills that theatre imparts may provide tools to help connect these communities in non-threatening and effective ways.

And what *did* I have to offer the Collaborative?

My discipline: theatre

A practising theatre artist, I have made my career over four decades performing and creating works for the stage; for the last twenty-five years I have taught in a university setting, during which time I continued to act, direct and write professionally. Involvement in my discipline has convinced me, whether through life-long immersion or by virtue of a natural predilection, that good storytelling makes for good pedagogy. I recognize the power of solid performance techniques to give lectures greater impact; the importance of coupling information (especially of the "harder," statistical kind) with human narratives to create more compelling classroom experiences; the strength of personal, dramatic stories to leave a more lasting impression. Theatre can aid in *making sense* of social, political and historical phenomena, by endowing them with tactile, visual, auditory and affective characteristics that leave a sensory, and therefore more lasting, impression in students' hearts and minds.[7]

With these beliefs in mind, I brought my theatrical experience to bear on our Climate Stories Collaborative workshops in 2017. The first two consisted of three major sections. First, as mentioned above, we presented a general overview of the origin of our idea and our concerns that led to our mission. In the second section—comprising the majority of the workshop time—we explored examples of ways art could engage with the science; in the final section—which proved to be the last fifteen minutes of Workshop Day Two—we opened discussion to all present about how we might want to move forward with the collaborative. England and I had decided that I would lead the second section, offering theatre as our primary example of how the arts might respond to environmental issues. For reasons I will explicate below, I had chosen to introduce and employ Boal's work: I had lectured on Boal in my Theatre for Social Change course—the student makeup for which consisting mainly of non-theatre majors—and thought his experiments would work best with people having little to no theatre experience.

I led the faculty participants in three activities. First, each shared personal encounters with the changing climate, or—if they had not personally experienced any such effects directly—memories of their engagement with the natural world. The recollections could be of anything, with the only stipulation that they focus on events that had affected them in a particularly emotional/visceral way. It is important to keep in mind that no other faculty participating in the workshop had much, if any, acting experience. Many in fact confessed various levels of performance anxiety. I was glad for their inexperience, as it served to demonstrate that much more powerfully the point I hoped to make with this particular activity (besides, all participated willingly). As each participant recounted a memory, I watched for the following four phenomena, examples of unconscious techniques that enhance the transmission of a story:

- sensorially charged details: the colour of a sweatshirt, the smell in the air, etc., infusing the recollection with specificity and vitality;
- gestures: bodily gestures and facial expressions the teller used the better to convey a particular moment, when words alone proved inadequate;
- dialogue: exchanges the teller recreated between different "characters" in the recollection;

■ present tense: times when the teller moved seamlessly (often unconsciously) into present tense in order to communicate the story with
more immediacy.

All participants in varying degrees employed most—and some all—of the
above techniques. I remarked upon this after everyone had finished their
recollection; they registered surprise but also recognition, and delight in
the variety of ways their colleagues had used gestures and improvised dialogue—had, in other words, used theatre to put their listeners in the time
and place of their story. This, I explained, had been my secret motive: to
show how these techniques reside in all of us. We know innately how to spin
a good yarn; therefore, I suggested, everyone in the room—whether artist
or scientist—could cultivate a more robust set of skills in order to communicate our experiences.[8]

The second exercise was built upon the first: I asked participants to
choose a partner and each tell their story to their partner once more. After
a few minutes of general din as everyone warmly repeated their anecdotes to
one another, I asked a few to summarize their partner's story. Then I asked
them to do it again, but *this* time to change the pronoun to first person, telling the story as if it were their own, which they did, after which I explained
that *this*—the telling of a story as if it were your own, moving beyond narrative into drama—lay at the heart of acting. The reader may in fact already
know these fundamental acting truths; however, since the larger purpose
here is to provide pedagogical tools to inspire conversations about environmental issues, thorough and explicit recapitulation may prove beneficial.
Basic tenets of storytelling and acting build the necessary foundation for
the workshop's third activity, which was a variation of Boal's Forum Theatre
(recounted more fully below, in Section 4).

The faculty responded positively to the games—although none had ever
encountered Boal before—and agreed enthusiastically that our college
should move forward in the creation of the Climate Stories Collaborative.
And the workshops themselves were not simply a "one-off" event but were
intended to be idea-incubators from which the faculty could draw inspiration as they created curriculum. Our desire was that faculty would retool
the activities and exercises we had introduced into workable class sessions
and assignments. Some colleagues also invited England and me to deliver
variations of our workshops to their classes. We decided further that we
should have an end-of-semester showcase, offering for display to a larger
public such artistic assignments as students had produced in class.

England and I continued to build on the support of the CFAA faculty,
visiting classes, and—with the help of many faculty and students, in particular, Jennie Carlisle, Chief Curator at the Smith Gallery on the ASU
campus—creating our first showcase, open to ASU faculty, staff, students
and the general public. The event featured art, sculpture, posters, crafts,
video, poetry, fiction and the presentation of short plays, all created and
performed by students (Figure 8.1).

Figure 8.1 Visitors attending the first Climate Stories Showcase (Winter, 2017). Over 200 students participated in creating the artwork, poetry, fiction, short plays presented at the showcase. (Photo credit: Shauna Caldwell.)

The Collaborative continued to grow. The first showcase in December 2017 displayed the work of over 200 students and drew a crowd of approximately 300 people; the next showcase, in April 2019, lasting a week and including related events such as a film night and a Forum Theatre event (based on exercises from our first workshop), saw the participation of over 1000 people. In its first year the collaborative included faculty from only the College of Fine and Applied Arts, but has since expanded and now enjoys university-wide participation.

I offer this summary of our Collaborative as testament to the overwhelmingly positive response to this sort of initiative, from faculty, students and the larger community. Clearly people want to talk about these things, and they desire guidance and leadership to help them discover the most effective ways of doing so.

I continued to find great satisfaction in the enthusiasm with which faculty and students embraced Boal's Forum Theatre. Its success led me to experiment with variations at my own university and to conduct a similar workshop for the Kennedy Center American College Theatre Festival (KCACTF) in Spartanburg, South Carolina, in February 2020. The success of this workshop, especially since I conducted it with students representing universities from all over the US southeast, proved that Boal's theatre games offer effective tools for initiating and improving climate change conversations.

Augusto Boal: Theatre of the Oppressed

Before describing the KCACTF workshop, some words about the ideas sub-tending Augusto Boal's Theatre of the Oppressed and his Forum Theatre are in order.

Boal's theatre was always an ambitious attempt to move beyond the tradi-tional Aristotelian notion of theatre as—in his words—a kind of "coercive system" designed to "bridle the individual" and pacify (by purging the dan-gerous emotions of fear and pity) the populace (*Theatre* 46). This kind of experience, which quells the spirit and dulls the spectator's engagement, runs counter to Boal's larger mission (greatly influenced by the work of Bertolt Brecht) to create a theatrical experience that stimulates the spec-tator "to transform his society, to engage in revolutionary action."[9] Over the course of his tenure with his theatre company, Teatro de Arena de São Paolo, Boal created a number of theatrical interventions—he called them "games," Forum Theatre being one of his most effective for transforma-tion (or at least for *rehearsing* transformation)—intended to give partici-pants tools for taking control of their own lives. A central component of his Forum Theatre was a new kind of spectator he named the "spect-actor" (*Games* 2), who, rather than remaining passively sitting in the audience watching action imitated on a stage—distant, removed, sedentary (much as one would watch film or television)—entered directly into the action being performed.[10] I speak more about the "spect-actor" when describing the KCACTF workshop below.

KCACTF

The first weekend of February 2020 I gave a workshop as part of the KCACTF Region IV conference, held that year in Spartanburg, South Car-olina. The conference is an annual festival for colleges and universities across the United States, an opportunity to come together to share their productions, take workshops and seminars, participate in staged readings and devised theatre projects.[11] My workshop, "Forum Theatre and Climate Change" transplanted what I had developed in the last two years as part of the Climate Stories Collaborative. It had been scheduled for the last day of the festival in the afternoon, so I was anticipating a slim turnout. Starting on a Tuesday, the KCACTF conference is a whirlwind of strenuous activity; students move into the weekend exhausted to the point of delirium. But the students attending numbered over thirty-five, a surprisingly large size for a workshop of this nature (the workshop's location, a medium-sized workout room, proved rather cramped for the number of participants); the generous attendance might indicate the hunger among theatre artists for activities addressing just these kinds of issues.

After the customary introductions, and my brief overview of Boal's work and his concepts surrounding the Theatre of the Oppressed, I presented them with the Yale Climate Opinion Maps, and asked how they accounted for the disparities. In other words, what did they think were reasons for our national avoidance in talking about climate change? They trotted out mostly unsurprising answers: climate change had become politically charged, causing them to avoid bringing it up and instigating heated political arguments. Many, they averred, assumed the science was still divided; others suggested it was because they didn't feel competent or knowledgeable enough to speak about the issues. Most in the room agreed, however, that more conversations should happen.

I then introduced them to Boal's Forum Theatre and explained that for the rest our session we would play this improvisation game. As the Festival serves theatre people, and the majority of attendees are either actors or have some acting experience, I skipped the first exercises built to introduce basic performance ideas to non-performers and jumped into *the rules of the game*. Most appeared excited by the context and anticipation of "improv games," and on my instructions broke into chatty, laughing groups of four to five people.

The Rules (following Boal): First, they were to come up with a scenario, which they would then perform before the entire room. I had explained to them that I wanted a scenario having something to do with issues of climate change, but that—for the purposes of our game—they need not have experienced the scenario themselves. They could draw from any stories (including, of course, their own) having to do with changing temperatures, mass extinctions, food or water insecurity, dramatic weather events, mass migrations and so forth. We all took a few moments to shout out different possibilities; then, the different groups split off and got to work creating their own scenarios. I had supplied pens, markers and paper for them to take notes and make outlines, or create makeshift props. I gave them between twenty and thirty minutes to work in their groups, during which time I floated from group to group to answer questions, help them fine-tune a scenario idea and remind them to get on their feet and rehearse a few times.

I established few additional rules for scenario-building and rehearsal:

Give It a Beginning, Middle, End: The scenario they created should observe the basic structure described by Aristotle to give their piece wholeness, clarity and impact (*The Poetics* 7.3). I articulated this rule after early workshops—in which participants consistently failed either to *begin* or *end* their scenarios—demonstrated my need to do so explicitly.

Include a Mistake: Boal explains the importance of including what he calls a "mistake" in the scenario, that is, a "political or social error" (*Games* 18–19), to which the participants, becoming spect-actors, can respond. It is these failures built into the scenario that can prompt engaged responses, inventiveness and collaboration in arriving at possible solutions, ways they—and ultimately, communities—can fix failures in real social or political systems.

Cast and Rehearse: It seemed self-explanatory that if groups were going to perform their scenarios they should fold enough time into their process to cast and rehearse. However, I learned after a few workshops that I needed to make this clear from the outset, with periodic reminders to the groups that they had a limited time before everyone was required to perform, so they would do well to get on their feet. Nonetheless, invariably a couple of groups had waited until the last minute to run through their scenario.[12]

After about thirty minutes (and a two-minute warning), groups began presenting their scenarios. The issues varied: one group chose to focus on extreme weather events (in this case because one of the participants had herself experienced a devastating hurricane); another created a non-realist "trial" in which the CEO of a coal company, a senator and a teenager wrangled with the personification of Mother Earth over who should decide how to use natural resources. Each of the five or six scenarios took between one and two minutes to perform. Everyone watched, laughed, applauded. The scenes were simple, straightforward and—except for the one depicting Mother Nature—grounded in a realist/naturalist style. They each, in varying degrees of subtlety and attention to verisimilitude, had folded in some error that led to an unsatisfactory conclusion (Figure 8.2).

The Spect-actor. We moved to the next, most crucial step of the Forum Theatre exercise. Each group repeated their scene, during which time members of the audience were invited to participate: they had been instructed to yell "Stop!" whenever anyone saw a moment in the scene they thought should go differently. But rather than explain what was wrong with the scene, they entered it directly, replacing the actor whose actions they disagreed with; then they improvised the moment as *they* thought it should proceed; or as they thought would most likely lead to the conclusion they preferred.

Figure 8.2 Students watching one of the Forum Theatre scenes, KCACTF, February, 2020. (Photo credit: Gina L. Grandi).

In the "Mother Earth" scenario, a spectator yelled "Stop!" then replaced the teenager, who had allowed the CEO to bribe the senator into passing legislation that would favour the coal company at the expense of the environment. The students made the switch; the new actor ad-libbed a delicious rant against the senator for acting against the welfare of his constituency, reminding him that she represented the future, and that if he did not listen to the young people they would be sure to vote him out in the next election. Thus our young spect-actor, reimagining the scenario to redress the "mistake" the group had purposely embedded in the script, created an ending that proved more satisfactory for Mother Earth and—while still remaining within the world of probability—more gratifying for the audience.[13]

It is with this action of the spect-actor that Boal's "Poetics of the Oppressed"— "to change the people ... into subjects, into actors, transformers of the dramatic action" (*Theatre 122*)—achieves its main objective. Opposing itself to the more traditional *telos* of theatre, the cathartic excitation of emotions that—according to Boal—deadens ambition and quells the desire to act, this movement of the spectator to spect-actor *stimulates* action by rehearsing it in the body of the spectator. They do not just watch actors perform actions; they perform the actions themselves, thereby helping them to practise physically actions they may then carry into real life. As Boal states, through such activities they will "be able to practice, to train for action, they will be able to act within the imaginary life of the theatre forum, so that afterwards, catalyzed, they can immediately apply this new energy to their real lives" (*Games* 246).

Another useful pedagogical by-product of this kind of theatre game is that—reflective of the name Boal has given it—it provides a playful, enjoyable forum for groups to discuss the issue of climate change. In the jubilant, imaginary space they have all agreed to enter, the words they say and the improvised decisions they make are non-binding, and can be unmade and remade by the next spect-actor. They can try on different responses, examine positions, adopt personae—and in so doing (with all due respect to Boal) may inadvertently gain empathy for people they had previously not fully understood, or even despised. As a collective they can wrangle over possible futures that, before playing, they had not imagined.

The workshop was a huge success. Students stayed well after the end of the session; many wanted to continue. A few remarked privately, and a little breathlessly, that it had been their favourite workshop of the festival. For me perhaps the most gratifying comment came from a student who runs an improv group in her hometown of Washington, DC: she said that Forum Theatre gave her so many ideas that she would fold into her improv sessions; ideas for making them *meaningful*, and not just mindless skits intended to entertain and get laughs. She loved the fact that one could use these kinds of improvisatory games to generate awareness. An activist herself, she was excited to have more tools in her activist toolkit; instruments besides a sign or megaphone to inspire action. She was already planning to use Forum Theatre in their next rehearsal.

Ultimately the workshop demonstrated how hungry our students are for just this sort of activity. Young people feel the imminence of massive, cataclysmic change and seek methods whereby they can participate in social and political machinery that can better prepare them, their families, and their communities for what is approaching. The young artists I work with are *not* scientists, but nonetheless want to bring their discipline into the conversation, employ their unique art form in service to creating a healthier, more sustainable future.

Forum and the future

One overwhelming, underrated fact, reinforced with every workshop I conduct, is that these activities are *fun.* The ludic nature of Boal's work, underscored in the title of his book *Games for Actors and Non-Actors,* creates an enjoyable, non-threatening environment wherein participants can play—if they approach the game in the proper spirit—without fear of reprisal, of judgment, of social stigma or political repercussions.

And because it is fun, Forum Theatre can also function as a great pedagogical tool. Research on the operation of our mirror neurons is producing exciting discoveries, and much about their behaviour and impact on human learning remains inconclusive (and receives vigorous debate). However, compelling evidence reveals a certain peculiarity: the mirror neurons fire in the brain and create synapses—initializing memory, the development of motor skills, emotional growth—irrespective of whether the body is watching, pretending, or actually performing an action. In other words, our brains are learning how to actually do things when we watch others do them, and—especially relevant in the case of Forum Theatre—when we imitate doing them. If this is so, by playing these games together, we are practising how to feel and understand each other, rehearsing how we want to act in the world; as Marco Iacoboni observes, our mirror neurons "demonstrate that we are wired for empathy, which should inspire us to shape our society and make it a better place to live" (Iacoboni 268).[14]

I began this chapter describing the "spiral of silence," in much of American society; its strange fear of discussing climate change. The theatrical activities introduced here start conversations. Moreover, they teach participants that such conversations are *possible.* The idea driving Boal's work is that his games *act as rehearsal for life.*[15] They prime participants to go out into the world, better prepared to have real conversations and to ready themselves for the next step, which is working courageously toward actual, real-world change. Forum Theatre gives them an entertaining opportunity to *rehearse* engagement, practise confronting leadership, and form brave words of resistance and solution in their mouths; they leave the classroom better equipped to actually engage, confront, resist and speak solutions.

And although Boal asserted that his theatre departed from Aristotelian formulations, I suggest that in a significant way Forum (and its variants) offers a much-needed *continuation* of them. If, as Aristotle maintained, drama is "the imitation of an *action,*" then it is action indeed, which is at the heart of Boal's mission.[16] We gather together, we play. We rehearse talking, debating, exercising our abilities to improvise, to think through ideas critically, to bounce them off one another in easy, enjoyable, non-threatening spaces. We prepare ourselves to move into real situations and to *act*. I find few better or more effective examples of the democratic principle at work.

Meanwhile, we must remember this: Boal's lifelong mission was to eradicate oppression. He travelled South and Central America, Europe and Africa in order to bring his theatre games to those communities sinking beneath the weight of economic, political, racial, religious and ethnic inequities. He hoped that through his games he would give these communities the necessary tools—found in theatrical and improvisational techniques— to fight and perhaps even overcome oppression.[17]

In this radically uncertain time, rocked by a changing climate, and besieged by powerful entities and systems bent on at best ignoring, at worst abetting, the effects of the environmental calamity which will dramatically affect every species on the planet, we are *all* the oppressed, struggling beneath the strain of a world that cannot sustain us. At the time of writing, an article in *The Guardian* has appeared asserting that we have six months to reverse, or at least avoid the worst effects of, climate change.[18] So... we have very little time to learn how to converse together. One way forward, as Boal suggests, is to start *playing* together.

Practically speaking ...

Boal's Forum Theatre model has great potential as a pedagogical tool in relation to any social justice issue, including climate disruption. Instructors are invited to seek out opportunities to train in this practice in order to be able to deploy it authentically and effectively.

Following is a review of the main rules for engaging in an act of Forum Theatre:

1. **Create a scenario.** Although the issue under examination in this chapter was climate change, the scenario can be an interrogation of any issue, as long as it is meaningful and can potentially impact the participants' lives. The scene can and should be short—under five minutes— but should nevertheless follow a recognizable structure (beginning, middle, end), and include enough characters to fully develop the conflict at the heart of the issue.
2. **Include a *mistake*.** Build into the scenario a "political or social error" (see page 11 above) that the spectators can discover, question, and reject in order to find solutions and imagine workable alternatives.
3. **Rehearse.** Once you have agreed on a scenario, cast it and rehearse as if it were a conventional, realistic scene (although there have been examples of more non-realist scenarios; see *Games* 22). Lines can be improvised, but actors should know their characters' objectives and obstacles, and should have a pretty clear idea of what they plan to say.

PLEASE NOTE: In my workshops I have rarely had the luxury of sufficient rehearsal time. Ideally, participants should know one other, be familiar with each other's work, and reserve for themselves at least thirty minutes for rehearsal.

4. **Perform.** The actors will perform their scenario once for their audience; afterward, the actors will ask if the scenario presented was satisfactory. If the actors had clearly embedded a mistake into the scene, the audience will likely respond negatively. The actors will perform the scene again exactly as they had before, explaining to the spectators that they are to participate more actively this time:

5. **Stop!** At any point during the second performance, spect-actors can yell "Stop!" when they see a character perform an action leading to social, political, and/or environmental disruption (whether a line, or gesture, or any other stage business). But rather than explaining their disagreement, the spectator must replace the actor performing the unacceptable action, entering into the scene itself, and perform the action in a way that (in their estimation) offers a more acceptable movement toward a solution to the issue under examination. This is the most crucial step in Boal's Forum Theatre, as it forces the spectator to become the spect-actor, and to rehearse playfully what can become effective actions in real-world circumstances. All participants are encouraged to repeat this step of the process as often as they like, replacing any characters and changing the scenario themselves until they collectively arrive at more satisfactory solutions to the problem being interrogated. This is Forum Theatre's heart: to give participants opportunities to practise *taking action*; as Boal reminds us, "the effect of the forum is all the more powerful if it is made entirely clear to the audience that if they don't change the world, no one will change it for them" (*Games* 21).

PLEASE NOTE: I must point out that one component of Forum Theatre I have not included here is the role of the Joker—a mediator of sorts, who explains the rules, helps spectators transition smoothly to the role of spect-actor, and generally keeps the ball rolling, as it were—because in my workshops I have taken on this role myself. But for a more thorough explanation of this and all the rules, see *Games* 18–21.

Notes

1 Naomi Klein persuasively asserts this perspective in *This Changes Everything*, although in a much more nuanced, economically focused way than perhaps her title implies.

2 Reasons for why this disparity is especially pronounced in the United States—ranging from the political to the economic—fall outside the purview of this chapter.

3 This is approximately a 3% drop from the previous year. See the Yale Climate Connections maps, https://climatecommunication.yale.edu/visualizations-data/ycom-us/. To gain insight into reasons for the discrepancy between the

number of people who recognize climate change to be real and the number of people who talk about it, see George Marshall's study from 2014.

4 Even so, the United States continues to rank among countries with the highest percentage of its population dismissive of climate disruption; see the Statista article from 2019, "Where Most Climate Change Deniers Live": https://www. statista.com/chart/19449/countries-with-biggest-share-of-climate-change-deniers/; or the recent summary from the Pew Research Center (also from 2019): https://www.pewresearch.org/fact-tank/2019/04/18/a-look-at-how-people-around-the-world-view-climate-change/

5 See the Yale Climate Connections site at https://www.yaleclimateconnections. org/.

6 The stated mission on their website: "The Climate Stories Collaborative is a transdisciplinary learning community aimed at growing the capacity of faculty and students to use a variety of creative media to tell compelling climate stories, including the stories of those who are affected by, and/or taking action to address, climate change." See the Climate Stories Collaborative, website at https://climatestories.appstate.edu/

7 One might argue that the purpose of the *ekkyklema* in ancient Greek dramas was to sear a visceral, unforgettable image into audiences as memory enhancement tool. See Frances Yates' (1966) magisterial work *The Art of Memory, passim,* for an exhaustive account of the theatre's mnemonic efficacy.

8 Aristotle speaks of this universal ability in *The Poetics* (see especially 4.1–6). But one need not go back two and a half millennia for authoritative confirmation; a more contemporary voice, David Mamet, asserts much the same thing—if a bit more humorously—in his extended essay *Three Uses of the Knife.*

9 Later Boal speaks unequivocally against empathy, which he calls "the most dangerous weapon in the entire arsenal of the theatre" (*Theatre* 113). Although I disagree, I understand this nuancing of the concept *vis á vis* his larger purpose of developing a new "Poetics of the Oppressed," which sets itself in opposition to Aristotle.

10 Technically, Boal introduces the idea of the "spect-actor" in *Theatre of the Oppressed* (especially when describing Forum Theatre, 139–42), but he never uses the term; rather, he refers to the "spectator-participant" or the "spectator-actor." It is not until *Games...* that he has made the elision to create "spect-actor," which while first mentioning it in the introduction (2), he does not fully describe until his longer explanation of the rules of Forum Theatre (19–21).

11 The festival comprises eight regions divided across the United States. For a full roster of the organization, consult their website: http://www.kcactf4.org/

12 Other standard components of Forum Theatre—for instance, the use of the "Joker"—have received less attention here in order to streamline my discussion; see Boal, *Games,* for a more thorough explanation of the "rules of the game," 18–21.

13 This is not to say everyone made realistic choices: one student yelled "Stop," then replaced the senator; he then proclaimed that he had had an awakening and would hereafter only pursue environmentally responsible legislation. His reversal of conscience, though fun for the audience, provided an instance of what Boal calls "magical thinking": while we may hope that those in positions of power would behave in a similar fashion, ultimately we cannot realistically hope for these convenient peripeties; rather, we can only hope to rehearse actions within our power to perform. See *Games...* 233.

14 Iacoboni's study, while chatty and accessible, offers a more sociological and quasi-philosophical perspective; see Rhonda Blair, *The Actor, Image, and Action* for an examination that more specifically intersects theatre with cognitive science (my thanks to Conrad Alexandrowicz for introducing me to Blair's work).

15 For more on these kinds of theatre, see Boal, *Theatre,* and *Games.*

16 Aristotle (1951), *The Poetics, passim.*
17 A number of extraordinary companies are continuing the work Boal began; see, for instance, the Cardboard Citizens in the United Kingdom, https://cardboardcitizens.org.uk/; the Theatre for Living in Canada, http://www.the-atreforliving.com/; and ActNow Theatre in Australia, https://en.wikipedia.org/wiki/ActNow_Theatre.
18 See Fiona Harvey's alarming if unambiguously titled article from *The Guardian*, 2020.

Works Cited

Aristotle. *The Poetics*, translated by S.H. Butcher, Dover, 1951.

Blair, Rhonda. *The Actor, Image, and Action: Acting and Cognitive Neuroscience*. Routledge, 2008.

Boal, Augusto. *Games for Actors and Non-Actors*, translated by Adrian Jackson, Routledge, 1992.

Boal, Augusto. *Theatre of the Oppressed*. Theatre Communications Group, 1979, p. 47.

Harvey, Fiona. "World Has Six Months to Avert Climate Crisis, Says Energy Expert." *The Guardian*, 18 June 2020.

Iacoboni, Marco. *Mirroring People*, Picador, 2009.

Klein, Naomi. *This Changes Everything: Capital vs. the Climate*, Simon & Schuster, 2014.

Mamet, David. *Three Uses of the Knife*, Vintage, 1998.

Marshall, George. *Don't Even Think About It: Why Our Brains Are Wired to Ignore Climate Change*. Bloomsbury, 2014.

Yates, Frances. *The Art of Memory*. U of Chicago P, 1966.

Part 3

Actor training

9 "Eco-atonement"

Performing the nonhuman

Conrad Alexandrowicz

One might think that the term "ecohubris" would have been in common parlance for some years now, given that it captures logically and usefully the radical nature of a global problem, one within which the climate crisis is subsumed. However, it seems to have been coined quite recently by Downing Cless, a contributor to *Readings in Ecology and Performance*:

> An extension of the ancient Greek term for an overabundance of pride or arrogance, *ecohubris* is an excessive zeal to control or dominate nature, acting without limits and with a sense of being above nature, as though one were a god.
>
> (160)

This encompasses the whole set of activities and processes, the excessive extraction, destruction, manufacturing, consumption, waste and pollution, that result from the largely unquestioned assumption of perpetual growth as the sole viable economic model.

The need to remedy the damages of ecohubris amounts to an ethical imperative for humans overall, working in any number of fields, and may be seen in the arrival in recent decades of a newly synthesized family of academic disciplines that may be collected under the name "environmental humanities." The contributors to *Teaching the Environmental Humanities*—a committee of twenty-three—include the following in the preface to their lengthy article:

> The traditional separation between those disciplines concerned with "nature" and those that examine "culture" has led to increasingly atomized science-based responses to environmental dilemmas. Work in EH seeks to develop and support alternative framings, approaches, and solutions that operate outside the dichotomized understandings of society and the environment which have underpinned diverse forms of colonialism, militarism, globalism, extractivism, and erasure. Living as we are in the midst of these violent global transformations, work in EH seeks to find modes of addressing environmental change that take seriously issues of justice, inequality, and oppression, and that value and support diversities of all kinds.
>
> (O'Gorman et al. 429)

At this point I must draw the reader's attention to this arresting caveat regarding conceptions of the non- and/or posthuman. Part one of the article by O'Gorman et al. is a global survey of post-secondary formations in environmental humanities, or other similarly disposed but differently named disciplinary areas. The passage pertaining to Africa contains the following:

> The 'post-humanities,' as a scholarly approach, has not found a great deal of traction in a context where, as one graduate put it, 'I've spent my whole life showing I am a human being not an animal; I cannot accept a post-humanism that wants me to become-animal.'
>
> (441)

It is of crucial importance to keep in mind such differences between the "global North" and "global South," functions of the long legacy of colonialism and racism. We say we are living in and responding to the effects of the Anthropocene Era, "a new phrase in the history of both humankind and of the Earth, when natural forces and human forces become intertwined, so that the fate of one determines the fate of the other" (Zalasiewicz et al. 2231). But one question we ought to ask is, "*whose* Anthropocene is it?" While all life-forms on earth—including all humans—must contend with the consequences of its effects, only certain human societies are responsible for its creation.[1]

How might theatre pedagogy situate itself within practices based on the principles articulated by our colleagues in environmental humanities? Evelyn O'Malley, lecturer in Drama at the University of Exeter, is one of many who thinks that theatre education needs to catch up with its neighbours in the academy: "A changing climate means that humanities teaching in higher education has to … cultivate an awareness that we can't answer certain questions without collaborating-with, becoming-with others—human and nonhuman" (69).

As we argue in the Introduction, we need as theatre pedagogues to identify and practise our own contributions to transcending the ancient gap in the Western philosophical and spiritual tradition between "Man/Nature" and "mind/body." The very fact that we have the words *body* and *mind* but lack a verbal signifier for the unified nature and function of the two—evidence for which has been revealed in current research in cognitive neuroscience[2]—is a reflection of what is a fundamental philosophical problem of Western culture, a model and paradigm that has long since become a globalized phenomenon. The assumption that we must "engage with the soma," that is, the body, means that we are currently "disengaged" from it, as though we could exist without our bodies. And, of course, this book is, in part, just one small component of a widespread project in the academy and among artists to remedy the profound and pervasive disjunction between *psyche* and *soma* in the post-Christian, post-Enlightenment West that is at the root of our predicament. We might use the term *bodymind*, as some have done,[3] but this hybrid only serves to re-inscribe the problem in the very attempt to address it.

Addressing ecohubris in the realm of performance, I propose, means exploring the possibilities of performing the nonhuman, a project that inevitably
exceeds the frame of realism, and entails engagement of the *bodymind*. This is
a problematic proposition indeed, given that most performance is all *about* the
human, which may explain why other disciplinary areas are so much further
ahead in terms of addressing the nonhuman, broadly considered. Pioneering
scholar Una Chaudhuri coined the term "species theatre" (*Research Theatre...*
45) to capture both the practice and pedagogy of performing the nonhuman, the logical antidote to what has been, as she states, the most anthropocentric of all the arts, its subject always and continually the human. As she
wrote, the theatre is "the least environmentally aware, most eco-alienated, and
nature-aversive of all the arts of the Western world" (*Stage Lives...* 102). And, as
I have noted elsewhere, even in the rare instances in recent dramatic writing
where animals have been portrayed—Wallace Shawn's *Grasses of a Thousand
Colors*, Mark Rigney's *Bears*, and Eric Coble's *My Barking Dog* are some examples—they have been explicitly and purposefully played in human terms; that
is, "the non-human is fixed at the level of text, rather than being taken up in
the psychophysical body of the actor" as impersonation (Alexandrowicz 178).

More and more, research into nonhuman animals, as well as other life-
forms, is revealing the degrees to which we are like other beings, and they,
in turn, are like us.[4] This is enormously useful, given the kinds of discrimination practised between human populations: "human difference, like that
of other species, appears against an overall background of kinship, forming a web of continuity and difference" (Plumwood 137). While the "nonhuman" may refer to animal, plant, mineral or elemental otherness, my main
concern here is with the animal variety, itself an enormous collection of life-
forms. The *Oxford English Dictionary* defines "animal" thus: "A living organism
which feeds on organic matter, typically having specialized sense organs and
a nervous system and able to respond rapidly to stimuli; any living creature,
including man" ("Animal"). I propose that grappling with the problems
of performing the *animal other* offers significant pedagogical rewards for
student actors, given the great range of expression of our animal relatives,
which embody varying degrees of continuity *with* and difference *from* us, a
matter I take up in more detail below.

It must be said that the title of this chapter describes something that is
logically impossible: we cannot *perform the nonhuman*, no matter the degree
to which our humanness is concealed by costumes, masks, make-up or pros-
theses. But a theatrical performance is a perceptual event, not only—or not
even—an illusionistic apparatus, and is activated and processed in the appropriate cortical centres in the brains of its spectators. Therefore, convincing
representations of animal behaviour and expression may be achieved by
means of actors' observations, the technical acumen involved in the use of
their voices and bodies, and what can be only poorly described as the "'incorporation of the other's essence."

The need to "perform the nonhuman" bespeaks an attitude that, apart
from prompting productive creative responses, allows for acts of *eco-reparation*

or *eco-atonement* for the destructive effects of ecohubris: "Reparation" invokes notions of damages that are payable for extensive destructive acts wrought by states or institutions, such as those proposed by African-American activists for the Atlantic slave trade.[5] The notion of atonement for sin is of course situated within religious belief and practice; this may be the most useful metaphor by means of which we may be inspired to attempt performing the nonhuman. The word "atonement" both means, and may be divided as, "at-one-ment," as the Oxford English Dictionary advises. Its first definition is the "condition of being *at one* with others; unity of feeling, harmony, concord, agreement," and its second, the "action of setting at one, or condition of being set at one, after discord or strife" ("Atonement"). Both of these meanings are entirely appropriate for my purposes here; that our efforts in performing the nonhuman are about achieving a radical recognition of and imaginative connection with the life-forms around us, after millennia of disjunction, itself the necessary psychic precondition for exploitation and destruction. And I am not necessarily arguing for instruction in a productive dramaturgy based on representing other-than-human life, but rather as a pedagogical pursuit with intrinsic value: students stand to benefit in any number of ways from performing aspects of other life-forms, regardless of whether such activity results in a performance object.

I first consider the philosophical positions that form the theoretical background to this project, paying attention to those that facilitate recognition and connection, and then consider some practical approaches for instructors. Performing the nonhuman in the manner I am advocating connects to notions of ritual observance, to meditation and to the spiritual dimension generally; another sign of the kind of radical transformation of our work as instructors that may be summoned in the face of the climate crisis. Such notions of the spiritual open a door to Indigenous belief systems and practices, considered by other writers in this volume. The link between imagination and spiritual practice is captured in an observation by renowned Haida artist Robert Davidson: "Whatever we can imagine, we can create" (*Haida Modern*).

In *Theatre and Ethics*, Nicholas Ridout notes the watershed in Western ethical philosophy formed by the Nazi Holocaust, in which millions of people were murdered on the basis of their ethnicity, disability, sexual orientation, or political or religious beliefs. The Lithuanian-Jewish philosopher Emmanuel Levinas was a prisoner for almost the entire duration of World War II, and his father and brothers, among others, were murdered by the SS.[6] He articulated a radically new concept of ethical obligation, oriented not to the realization of the self, core to the Western philosophical project thus far, but in recognition of and care for the "Other" (Ridout 52). Levinas' compelling, poetic notion of the Other is decidedly religious in its inflection:

> Responsibility for the Other, for the naked face of the first individual to come along. A responsibility that goes beyond what I may or may not have done to the Other or whatever acts I may or may not have committed.
>
> (Levinas 83)

But while articulated within the *humanist* tradition, it offers an inspiring model for the purposes of connecting to the *nonhuman*. Ridout acknowledges this possibility, albeit indirectly, noting that "Levinas seeks to remove the human subject from its former place at the centre of the world" (53), and thus "the theatre of moral instruction gives way to performance as ethical practice" (54). This is thoroughly germane to the project of performing the nonhuman.

The works of philosopher Gilles Deleuze, in particular his collaborations with political activist and radical psychotherapist Félix Guattari, are of signal importance in the playing of otherness. As my co-editor David Fancy has observed, their "rejection of normative notions of bounded identity and subjectivity in favour of something more expansive, multiple and, ultimately, more playful … resonates with unexpected potential" (93). (I address the practical use of their work in some detail below.)

De-centring the human subject leads to the terrain mapped by various forms of ecofeminism, whose positions are articulated with particular thoroughness and precision by philosopher Val Plumwood. The core of her argument—and those of the movement's other founding thinkers, such as Ariel Salleh, Freya Matthews and Kate Rigby (Stevens et al.)—is radical and all-encompassing: foundational elements of Western philosophy have enabled and justified the pillaging and pollution of the earth's resources, the oppression of women, the exploitation of labour by capital and the enslavement of racialized populations. Thus, disastrous effects in both the biosphere and in the "sociosphere" (*Feminism* 193) have a common cause. Plumwood argues that such relationships of power and dominance are fused in the notion of "mastery," typically and historically the prerogative of elite white males, and that this set of linked concepts can be traced to the very wellspring of Western thought. As she notes, "Aristotle, in a notorious passage in the *Politics* justifying slavery, links together the dualisms arising from human domination of nature, male domination over females, the master's domination over the slave and reason's domination of the body and emotions" (46). While employing different terms, her position aligns with that of Bruno Latour, that the reason/nature dualism is primary, giving rise to, confirming and supporting the others (46–47), and that this set of linked dualisms has persisted through the mainstream of Western philosophical discourse, from the Platonic and Aristotelian, to Christian rationalist and Cartesian rationalist formations (70). Enlightenment philosophy, exemplified by Descartes, accompanied the age of imperialism, of plunder and destruction on a massive scale, revealing in practice theoretical potentials articulated roughly fifteen hundred years earlier in Ancient Greece. Plumwood states this succinctly: "A licence for the annexation of nature is provided by Cartesian mind/nature dualism, the close associate of Cartesian mind/body dualism" (111). It can be dizzying trying to connect personal experience to such universalizing schemes: how can the privileging of the text over the body in our theatrical tradition be linked to centuries of European imperialism? But Plumwood's analysis—and that of her contemporaries—is cogent, based on volumes of evidence, and offers potent fuel for the topic of this chapter in particular, and the overall project of this book.

Applied philosophy

As Peta Tait has observed, speaking of the Australian context, "Aboriginal knowledge involves a fundamental philosophical belief that humans have been physically made by the land and are co-descendants with other species" (361). How can we begin to move beyond the ancient, pervasive and potent ideas of division and exclusion to which we are heir in the West, and towards such notions of connectedness, marked by interdependency, responsibility and co-descendance? According to Plumwood, resolving the linked dualisms of "mind/body" and "human/nature" would entail establishing relationships between the terms of each pair that are both non-hierarchical (60) and non-reductionist (123): we need to recognize and greet "earth others," as Plumwood calls them (137), as peers, both like and unlike us, and equipped with distinct features, not reducible to type. Abiding by these notions would address the caveat noted above, regarding the logical impossibility of "playing the nonhuman": "Overcoming the dualistic dynamic requires recognition of both continuity and difference; this means acknowledging the other as neither alien to and discontinuous from self nor assimilated to or an extension of self" (6). This is crucial, as it requires us to be rigorous in our attempts to represent various kinds of animal kin, avoiding demonizing, caricaturing or anthropomorphizing them.

Plumwood speaks from a unique position in terms of understanding the other-than-human: in 1985, while canoeing in the renowned Kakadu National Park in Australia's Northern Territory, she was attacked by an estuarine crocodile, dragged underwater and subjected three times to the legendary "death roll" the animal uses to subdue and drown its prey. She managed by some miracle to escape—or rather, *was released*—and, although gravely injured, was able to hike to a location where she was eventually found by a park ranger. She was hospitalized in intensive care for a month and underwent extensive skin grafts. In writing about this traumatic experience, she explored the human response to the unthinkable condition of becoming the prey of another animal, her narrative of human selfhood erased in "becoming meat." As she notes, in much Aboriginal thinking, perhaps in common with the experience of other hunter-gatherer peoples, now and in our shared past, "animals, plants and humans, are seen as sharing a common life force, and many interchanges of form between human and animal are conceived. In the West, the human is set apart from nature as radically other" ("Human" 34), and therefore becoming food for another being is obscene and inconceivable, in particular in a culture that fears and denies death itself. In our human supremacist world-view, we are the master predator, and therefore being killed and eaten by a crocodile— or other apex predator—compounds the horrors; "the decomposition of the body with active animal triumph over the human species" (ibid.). It is for this reason that any attack on a human by, for example, a grizzly bear or cougar in my part of the world, British Columbia, results in a concerted effort on the part of "conservation officers" to hunt down and kill the individual animal responsible for the attack, if possible, and if not, given that it has usually vanished from the scene before their arrival, any representative of the offending species; it becomes a matter of species-supremacist revenge.

Plumwood's terrifying experience has valuable lessons to impart to all of us, one of which is the requirement that we not sentimentalize or anthropomorphize the animal other. As part of our ethical practice as instructors we must supervise students' activities of observation and representation with awareness of the various potentials of other animals, some of which can be lethal to us. Part of playing the nonhuman entails an attitude of the kind of deep respect that we observe in Indigenous knowledges and practices and commands us to transform the ways we think about our world, so that, as Plumwood proposes, we "discover the body in the mind, the mind in the animal, the body as the site of cultural inscription, nature as creative other" (*Feminism* 124).

Playing the animal other

There are many approaches to the playing of the nonhuman, and I consider a number of approaches, including the pedagogy of Jacques Lecoq, the uses of both Viewpoints- and dance-based improvisation, and the inspirations of Indigenous cultural practices. I propose that such methods stand as a guide for acting instructors who are committed to developing radically new approaches to teaching student actors. Many of us as faculty members forget, once we obtain our doctoral or master's degrees, about the possibilities for becoming students again, for ongoing re-education and renewal of our capacities as teachers and researchers. There is perhaps no more pressing impetus for learning new skills with which to teach than the environmental emergency. As Greta Thunberg has proclaimed, "[w]e now need to change practically everything. We now need a whole new way of thinking" (73).

The pedagogy of Jacques Lecoq is full of explorations of the "imagining body" into the realm of the nonhuman; in various phases of the curriculum students explore performing as animal, substance and weather phenomenon, as I and many other theatre artists and instructors know from personal experience. Alumna Martha Ross, a Toronto-based theatre artist, director, and teacher, recalls the plethora of nonhuman inspirations undertaken at the Ecole Jacques Lecoq:

> I remember that we worked a lot on the four elements, and then colours, and then dynamics of light, and then man-made materials, like paper uncrumpling, or something more forceful like fireworks or

popcorn. And then animals: I don't remember exactly which ones but we did a lot of creatures: insects and snakes and on up the food chain ... birds ... primates ... I think you can identify with absolutely anything that moves and find its dynamic.

(Conversation with Martha Ross, June 19, 2017)

Part of my post-dance training entailed a course in the Neutral Mask with Philippe Gaulier in London, who was first a student of Lecoq and then his collaborator. His workshops attract student and professional actors from around the world, and the study of this mask, essentially invented by Lecoq, has great potential for the playing of various registers of otherness. (I recall one particularly memorable improvisation in which we had to play hydrochloric acid.) I have incorporated the basic method I experienced as a student with Gaulier into my work teaching "movement for actors" at the University of Victoria. Students improvise in the mask in small groups while the rest of the class observes. Supported by the use of appropriate selections of music, students move and gesture as fire, the colour red, one of the Seven Deadly Sins, chickens, sheep, hummingbirds, a thunderstorm—almost any source may serve. Then the music is silenced, and the actors freeze while classmates who have been observing hasten to remove their masks. They are then called upon in turn to speak or sing in the mode of the source. Because one's memory can be undone in the heightened psychophysical states induced in the improvisations I recommend they begin with speeches they know very well from their text classes. But after some sessions of such improvs, I ask them to improvise speech in the moment, out of which characters emerge that are then fixed in text as monologues; these are among the many original assignments students create in my classes. Simon Murray notes that the representation of animals in Lecoq's pedagogy "is not a question of imitation and mimicry, but of 'becoming' or 'being' the animal or material in question" (75), but that this expression is then adapted to the playing of human characters. While I fail to see how "imitation" is not involved in the attempt to "become an animal," I take the point about the transposition of animal qualities to human expression; that is, in this context, playing the other-than-human was and is typically deployed in order to develop increased range in the playing of *human* characters. This, I assume, was also the point of both Lecoq's and Gaulier's use of the neutral mask, and has remained the pedagogical objective of many instructors who have employed this mask as a teaching tool, myself included. And indeed, the realist tradition is full of examples of finding the "animal emblem" of a character: Steve Vineberg, in *Willy Loman and the Method*, reminds us "that [Lee J.] Cobb's preparation for the role [of Willy] relied heavily on a Method improvisation known as 'the animal exercise,' and the animal he chose to study for Willy was the elephant" (157). But, as I have argued, in the context of climate activism, "becoming animal" is an end in itself; indeed, it may have acquired the status of an ethical imperative. The objective, therefore, would be to play the elephant, as much a tribute to the nonhuman, and a mode of psychic healing, as a task of impersonation.

Research Theatre, Climate Change, and the Ecocide Project, co-written by Una Chaudhuri and Shonni Enelow, is at once a theoretical work about performance and climate change, and a practical handbook developed from a series of workshops with actors, directors and dramaturges—the two authors. And it contains the text of the play that emerged from the work, *Carla and Lewis*, written by Enelow. The authors offer an intriguing, at times puzzling, account of "becoming-animal," inspired by *A Thousand Plateaus: Capitalism and Schizophrenia* by Deleuze and Guattari. Chaudhuri et al. struggled with how to render in theatrical terms their observation that "[t]here is a reality to becoming-animal, even though one does not in reality become an animal" (*Research Theatre* 11). In what might this reality consist? As Chaudhuri recalls, "the most challenging of Deleuzian definitions ... was the idea that becoming is antithetical to imitation" (14). She quotes directly from *A Thousand Plateaus*: "We fall into a false alternative if we say that you either imitate or you are. What is real is the becoming itself, the block of becoming, not the supposedly fixed terms through which it passes" (ibid.). I find this both confusing and opaque: what other mode of "becoming" is available to us if not via a process of *imitation*? Surely, given that we cannot *be* animals we can only *represent*—that is, *imitate*, certainly not a pejorative term—their characteristics, via the deployment of carefully acquired vocal and physical techniques; yet, as the writers object, "we are not interested in characteristics" (ibid.). Or is this just semantic haggling?

As the reader will have noted, co-editor and co-contributor Dr. Fancy has an entirely different and—it must be said—much more expert engagement with the work of Deleuze and Gauttari. I leave the reader to make her/his own evaluation of Deleuze's and Guattari's strategies for "becoming-animal." It is important to remember that neither was a theatre practitioner or instructor, and that both operated in an extremely rarefied cognitive realm. But the point is that all of us addressing the topic in this volume are attempting to arrive at essentially the same destination; we are all rowing the same boat, even though we may be using different oars.

Perhaps the most useful dimension of this work lies in its extravagant poetic invention, particularly the metaphor of animal becoming as something that is external to and arrives in the body of the performer as possession, seizure or contagion, as Chaudhuri writes (ibid.). (The latter term has acquired a whole new weight in the present circumstances.) As noted above, this invites comparisons with the various notions of the numinous, of spiritual belief and of Indigenous cultural practices. But, conversely, it also invites us to consider how we actually learn and reproduce movement patterns in a coordinated way.

The attempt at "becoming-animal" does indeed begin with something external to the performer, given that it is properly based in observation, either direct or mediatized—those with easy access to other animals have a distinct advantage—and subsequently in the "mind's eye," that part of the brain where we can create moving images. Visual stimuli are received and processed in the occipital lobe at the back of the brain, then sent to various other of its parts, including, of course, the motor cortex at the rear of the frontal lobe. But "so many different structures in the brain are involved in motor functions that ... practically the entire brain contributes to body movements" ("Motor Cortex"). Movement material must be learned in order to be reproduced, and in this regard one must consider the extraordinary functions of the cerebellum:

> It stores learned sequences of movements, it participates in fine tuning and co-ordination of movements produced elsewhere in the brain, and it integrates all of these things to produce movements so fluid and harmonious that we are not even aware of them.
>
> (ibid.)

Cerebellar functions seem to be key to creating the impression of ease and wholeness that is the mark of a successful performance, and the goal to which training is largely directed. "Talent is co-ordination," as a teacher of mine once remarked. Chaudhuri describes the very demanding procedures of the workshops undertaken as part of the project in Research Theatre, in which actors attempted to shift in improvisations from human to animal and back again, with various stops in between. In this "Transformation Etude" Director Fritz Ertl assigned actors the following sequence of activities:

I am myself, a human animal.
I transform slowly into an animal.
I bump back to my human self.
I transform to a cartoon of the same animal. I speak.
The real animal rises from within to replace the cartoon.
A new (second) animal overtakes the original animal from without.
The original (first) animal emerges to co-exist with the new animal.
The cartoon animal emerges to co-exist with the new and original animal.
Your human self emerges to co-exist with the cartoon, original, and new animal.
Your conglomerate self becomes molecule. (*Research Theatre* 16)

Consider the multipartite yet fully integrated brain activity required to produce all these shifts and combinations! Such an assignment would be enormously difficult even for highly trained actors experienced in varieties of physical theatre, demanding tremendous skill, precision and concentration. While certainly inspiring, exercises such as these would require considerable adjustment for use in a classroom setting.

Chaudhuri warns of the dangers of retreating into "a sentimental and personalized quest for one's 'inner animal,' ... the one enshrined in habit, personal narrative, and collections" (14). I propose that Lecoq's methods both provide for the kind of productive "seizure or contagion" about which Chaudhuri speculates, and also protect the actor from lapsing into indulgence, which often leads to caricature, a failing I have witnessed countless times over many years of instructing actors in the use of the neutral mask to represent animal sources. I recall an experience as an audience member that supports this claim:

One notable Canadian graduate of Lecoq's visionary pedagogy is theatre-maker Dean Gilmour, who in 1980 co-founded his company, Theatre Smith-Gilmour, with his wife Michele Smith, herself a Lecoq alumna. The company has staged scores of productions, many of them original plays. One of these was a show that unfolded in a series of versions, retaining the title *Chekhov's Shorts*, based on Chekhov's short stories; performed a number of times in Toronto, it also toured internationally, winning numerous awards. It included the performance of a dog by one of the female actors that remains in my memory as one of the most arresting, convincing and powerful theatrical evocations of the other-than-human I have seen. I recall that it was a poignant and poetic tribute to canine life, portrayed in human flesh, rather than a trick or a cartoon. Such, in my view, is the power of Lecoq's pedagogy, and its potential for performers and spectators, grounded as it is in rigorous technique. As I have argued elsewhere, "the transformations we require may be obtained by circling back on decades of physical theatre pedagogies with new insights in hand regarding how to situate them, both as pedagogical devices and as methods of generating content" (Alexandrowicz 183).

Improvisations that arise from dance-based practices provide another route to playing the nonhuman. Dance artist/scholar Christine Bellerose has reflected on the experience of participating in a workshop in which the instruction was to move as though one had wings. While being able to accomplish and sustain the "imagined reality" of being winged at the experiential level, she of course remained aware that an observer would likely only be able to note certain subtle differences in her body: "The kinesthetic, somatic, proprioceptive penetration inward, and the visual-kinetic, imaginative reach outward, opens up my rib-cage, allows more air than usual to flow in and out of my lungs, weighs my body to the ground while lifting me up" (58). She calls this "somatic architecture" (ibid.), and it may be a useful term to describe the imaginatively produced structure of otherness, potentially involving voice coordinated with movement, that the performer finds through the complex neural processes alluded to above. Some animals, such as the dog who appeared in *Chekhov's Shorts*, are much easier for humans to represent than others; winged beings are of a different order of difficulty. But the pursuit of the embodied experience of otherness offers multiple rewards for the student, as performer, as individual and as citizen, regardless of the degree to which a credible representation is perceived by the observer. And the humility that is required to engage in pursuit of this essentially impossible goal connects with the spiritual dimension of performance training as an act of eco-atonement.

Bellerose notes that as a dancer she "understands anatomical details involved in producing movement ... [and] has been trained to project herself into the experience of a body-self tending toward otherness" (58–59). This kind of practice—or rather its particular points of departure, gathered under the sign "dance"—may be beyond the capacity of many acting instructors, as well as their students, unless they have had the years of training that result in technique that can be "thrown away" (It was often said that "it takes ten years to build a dancer.") Over many years of working in the area of "movement for actors," I have noted both the benefits and limitations of technical training in dance for students; the former may be outweighed by the latter: Virtuosic vocabulary undertaken for its own sake may overcome the capacity of the actor to find idiosyncratic movement expression, their *movement-print*, analogous to their fingerprints, if you will. Further, the cornerstone of most dance training is ballet, which, in most young people's limited experience with the art form, tends to produce a polite restraint in movement that is antithetical to the kind of kinesthetic risk-taking one seeks to instill in student actors.

"Physical" approaches to actor training—as opposed to those that are more clearly cognitive in emphasis—have been gaining ground in mainstream North American post-secondary programmes in recent decades, as I have discussed elsewhere. (*Acting Queer…* Chapter five, "Gendered Movement and 'Physical' Acting.") The method detailed in *The Viewpoints Book: A Practical Guide to Viewpoints and Composition*, by Ann Bogart and Tina Landau, has proved one of the most popular since it appeared in 2005. The nine viewpoints of Tempo, Duration, Kinesthetic response, Repetition, Spatial relationship, Topography, Shape, Gesture and Architecture together amount to a kind of improvisation machine, through which can be run any source element: an idea, a theme, a word, a line of text, a scene, a play, a particular event—or an animal, or some other nonhuman entity. The viewpoint of Architecture, considered broadly as *environment*, offers perhaps the most useful lens in terms of responding to animal imagery. I, like many instructors, have supervised students' use of the Viewpoints method of collective devising as a route to theatrical expression, but it can equally be thought of as a dance-based practice, one that radically democratizes the practice of that art form, contiguous and overlapping with theatre. Through a conceptual shift an "open Viewpoints session" (Bogart and Landau 54) may generate the kind of abstract vocabularies we conventionally associate with dance improvisation, in which anyone can participate, regardless of physical ability.

As I have also argued (Alexandrowicz 185–188), a rich source of inspiration for playing the nonhuman may be found in various Indigenous cultural practices, including performance, where the sense of connection to and interchangeability with nonhuman life has been part of lived experience for millennia. All of our philosophical struggles with the human/nature/mind/body trapezoid, analysed and dismantled with noteworthy rigour by Plumwood and many others, seem to disappear in the light of Indigenous ways of knowing, which have also been ways of *being*. Of the many scholarly treatments of this topic I note Peta Tait's *Enveloping the Nonhuman: Australian Aboriginal Performance*, which considers samples of contemporary work by "Australian Aboriginal and Torres Strait Island performers that draw on modes of movement and storytelling in which there is continuous exchange between human and animal identities and environments" (348). Indigenous performance practices may be seen to have a salutary influence in various contexts in dominant cultures, in the light of the demonstrated failures of Western thought—and therefore its practices—in relation to the bio-world of which we are a part, and on which we depend for survival: "The embodied practices of Aboriginal culture that enact connection with the nonhuman surroundings have the potential to enhance Western cultural understanding" (360). But there is an acute irony that must be noted here: the relative availability of Indigenous belief systems and practices to those in settler societies has only been made possible by the effects of invasion, dispossession and disruption.

As Tait notes, the contemporary works she analyses have their origins in community and ceremony (351), and were practised with differing inflections by Aboriginal nations that were more or less geographically intact before European invasion. And stories depicting particular human/other–animal relationships, in the cosmology of "the Dreaming," belonged to certain tribes and clans, were handed down through generations and were not to be shared with others, never mind being sold as objects of cultural consumption (348).

In the course of centuries of disruption, forced relocation and migration, Aboriginal peoples have been mixed with and among each other, as well as with those of European descent, and have developed a range of hybrid cultural expressions, acquiring training, for example, in Western dance forms (352). This hybridization and urbanization have situated Aboriginal world-views and practices within the cultural marketplace, where they may be appreciated—or appropriated—by festival producers, artists and academics. (This process has, of course, unfolded all over the post-colonial world and has produced all the complications of interculturalism, amplified in recent decades by the effects of globalization, about which much has been written. See, for example, *The Intercultural Performance Reader*, ed. Patrice Pavis, Psychology Press, 1996).

These same considerations, of performance as clan-specific sacred ritual, and of the disorienting effects of disruption, also apply to Indigenous peoples of North America: many have been uprooted, transplanted, mixed with their counterparts from disparate geographical locations, their practices blended with and adapted to Western forms. Ric Knowles, writing about the Living Ritual International Indigenous Performing Arts Festival, held in Toronto in July 2017, notes debates among artists and in Indigenous scholarship regarding "Nation-specific" vs. "trans-Indigenous" approaches (83). He cites Chadwick Allen, of Chickasaw ancestry, who, arguing for use of the term "trans-Indigenous," acknowledges "the mobility and multiple interactions of Indigenous peoples, cultures, histories and texts" (xiv).

Exploring the possibilities for such Indigenous guidance in our field as acting instructors is appropriate, given that, among the ninety-four calls for action articulated by the Truth and Reconciliation Commission, issued in 2015, one finds the following:

> We call upon the federal, provincial, and territorial governments, in consultation and collaboration with Survivors, Aboriginal peoples, and educators, to … [p]rovide the necessary funding to post-secondary institutions to educate teachers on how to integrate Indigenous knowledge and teaching methods into classrooms.
>
> (*Truth* 7)

However, while we engage in such processes of consultation, with the aim to implement their findings in practice, we must beware of the perils of both cultural misunderstanding and appropriation.

What this might mean is that we find our own ways of performing the nonhuman, searching back into our own blended cultural inheritances, perhaps under the guidance of Indigenous artists and/or elders. We have barely begun to attempt such curricular transformations, and therefore questions abound regarding how they might be meaningfully undertaken. But it is worth noting that there is considerable institutional support for such measures, at least in some quarters. For example, "indigenization" is one of six core components of the Strategic Plan of my university.[7]

I conclude with these words of Val Plumwood:

> The reason/nature story has been the master story of western culture. It is a story which has spoken mainly of conquest and control, of capture and use, of destruction and incorporation. This story is now a disabling story. Unless we change it, some of those now young may know what it is to live amid the ruins of a civilisation on a ruined planet. … If we are to survive into a liveable future, we must take into our own hands the power to create, restore and explore different stories, with new main characters, better plots, and at least the possibilities of some happy endings.
>
> (*Feminism* 196)

Practically speaking…

- If you have training in Lecoq's methods, consider how you might revive them in the areas of playing the nonhuman. If not, find a Lecoq-trained instructor who offers courses in neutral mask or linked practices.
- Learn the methods of improvisation and composition in *The Viewpoints Book*, which were designed in a spirit of democratic inclusiveness. Viewpoints improvisation also lends itself well to improvising in outdoor environments. Perhaps our acting students should do much more of their training outdoors, also a relatively safer environment, in terms of the pandemic, than studios and classrooms.
- In the spirit of (re)conciliation with Indigenous peoples, explore the possibilities for contact and consultation with artists and/or elders in your community. Our efforts in this regard tend to begin and end with the conventional land acknowledgements before public events; how

can we as instructors engage in meaningful interaction with our Indigenous neighbours and colleagues?

Notes

1 For arguments concerning "Anthropocene" and alternative terms, see Donna Haraway's *Staying with the Trouble: Making Kin in the Chthulucene*, Duke UP, 2016.
2 The truly unitary nature of our being is being continually demonstrated in the latest research in cognitive neuroscience, a rich source of which for theatre practitioners is Rhonda Blair's *The Actor, Image, and Action: Acting and Cognitive Neuroscience*. London and New York: Routledge, 2007.
3 Or "mindbody," as the case may be: see the work of philosopher William Poteat: https://divinity.yale.edu/news/my-mindbodily-being-celebrating-william-h-poteat-44-bd (accessed 7 June 7, 2020).
4 For astonishing revelations about the capacities of tree communities the reader should consider Peter Wohlleben's *The Hidden Life of Trees: What They Feel, How They Communicate*. See https://www.smithsonianmag.com/science-nature/the-whispering-trees-180968084/ (accessed 8 June 2020).
5 See, for example, https://www.bbc.com/news/world-us-canada-48665802 (accessed 16 May 2020).
6 See https://plato.stanford.edu/entries/levinas/#LifeCare (accessed 25 May 2020).
7 See this component of the Strategic Framework of the University of Victoria. (Co-contributor and colleague Alexandra Kovacs explores other elements of this document in some detail) https://www.uvic.ca/strategicframework/priorities/reconciliation/index.php (accessed 15 June 2020).

Works Cited

Alexandrowicz, Conrad. *Acting Queer: Gender Dissidence and the Subversion of Realism*. Palgrave, 2020.

Allen, Chadwick. *Trans- Indigenous Methodologies for Global Native Literary Studies*. U of Minnesota, 2012.

Animal. *Oxford English Dictionary*. 2010. https://www-oed-com.ezproxy.library.uvic.ca/view/Entry/273779?rskey=EC95LC&result=1#eid (accessed 7 June 2020).

Atonement. *Oxford English Dictionary*. 1989. https://www-oed-com.ezproxy.library.uvic.ca/view/Entry/12599?redirectedFrom=atonement#eid (accessed 23 April 2020).

Bogart, Ann and Tina Landau. *The Viewpoints Book: A Practical Guide to Viewpoints and Composition*. Theatre Communications Group, 2005.

Bellerose, Christine. "On the Lived, Imagined Body: A Phenomenological Praxis of a Somatic Architecture." *Phenomenology & Practice*, vol. 12, no. 1, 2018, pp. 57–71.

Chaudhuri, Una. *The Stage Lives of Animals: Zooësis and Performance*. Routledge, 2017.

Chaudhuri, Una, and Shonni Enelow. *Research Theatre, Climate Change, and the Ecocide Project: A Casebook*. Palgrave Macmillan, 2014.

Cless, Downing. "Ecodirecting Canonical Plays." In *Readings in Performance and Ecology*, edited by Wendy Arons and Theresa J. May. New York, Palgrave Macmillan, 2012.

Fancy, David. "Difference, Bodies, Desire: The Collaborative Thought of Gilles Deleuze and Félix Guattari." *Science Fiction Film and Television*, vol. 3, no. 1, 2010, pp. 93–106.

Haida Modern: The Art and Activism of Robert Davidson. Directed by Charles Wilkinson, performances by Robert Davidson and others, Shore Films, Optic Nerve Films, Knowledge Networks, 2019.

Knowles, Ric. "Because It's Ritual, and We're Living: Living Ritual International Performing Arts Festival." *Canadian Theatre Review*, vol. 174, Spring 2018, pp. 83–88.

Latour, Bruno. *We Have Never Been Modern*, translated by Catherine Porter. Harvard UP, 1993.

Levinas, Emmanuel. *The Levinas Reader*, edited by Séan Hand. Blackwell, 1989.

Motor Cortex. *The Brain from Top to Bottom*. 2002. https://thebrain.mcgill.ca/flash/i/i_06/i_06_cr/i_06_cr_mou/i_06_cr_mou.html (accessed 12 June 2020).

Murray, Simon. *Jacques Lecoq*. Routledge, 2003.

O'Gorman, Emily, Thom van Dooren, Ursula Münster, and Dolly Jørgensen. "Teaching in the Environmental Humanities." *Environmental Humanities*, vol. 11, no. 2, November, 2019, p. 429.

O'Malley, Evelyn. "Theatre for a Changing Climate: A Lecturer's Portfolio." In *Posthumanism and Higher Education: Reimagining Pedagogy, Practice and Research*, edited by A. Bailey and C.A. Taylor. Palgrave, 2019.

Plumwood, Val. *Feminism and the Mastery of Nature*. Routledge, 1993.

Plumwood, Val. "Human Vulnerability and the Experience of Being Prey." *Quadrant*, vol. 39, no. 3, 1995, pp. 29–34.

Ridout, Nicholas. *Theatre and Ethics*. Palgrave Macmillan, 2009.

Stevens, Lara, Peta Tait, and Denise Varney. "Introduction: 'Street Fighters and Philosophers': Traversing Ecofeminisms." In *Feminist Ecologies: Changing Environments in the Anthropocene*, edited by Lara Stevens, Peta Tait and Denise Varney. Palgrave Macmillan, 2018.

Tait, Peta. "Enveloping the Nonhuman: Australian Aboriginal Performance." *Theatre Journal*, vol. 71, no. 3, 2019, pp. 347–363.

Truth and Reconciliation Commission of Canada: Calls to Action. "Truth and Reconciliation Commission of Canada." 2015. http://trc.ca/assets/pdf/Calls_to_Action_English2.pdf (accessed 19 February 2020).

Thunberg, Greta. *No One Is Too Small to Make a Difference*. Penguin, 2019.

Vineberg, Steve. "Willy Loman and the Method." *Journal of Dramatic Theory and Criticism*, vol. 1, no. 2, 1987, pp. 151–162.

Zalasiewicz, Jan, Mark Williams, Will Steffen, Paul Crutzen. "The New World of the Anthropocene." *Environmental Science and Technology*, vol. 44, no. 7, 2010, p. 2231.

10 The actor as geoartist

David Fancy

All performance is geoperformance. All acting is geoacting. All theatrical artistry is geoartistry. Geoartistry is not restricted to human artistic activity. In fact, the premise of geoartistry is that many phenomena with a range of different types of sentience can not only produce but also experience artistic activity. What does this mean? On a basic level, it means that the wider context in which human acting occurs is the context of the planet: of all the entities, phenomena and topographies constitutive of the Earth. The implications of foregrounding such a position are simultaneously obvious and radical. A vicious and globally dominant anthropocentrism—fuelled by the excoriating capitalist economic practices of objectification, extraction and pollution of this same Earth—is immediately called into question. And yet, being attendant to interconnections and mutual influence between human and other-than-human entities is only part of the solution of decentralizing anthropocentric narcissism. Humans have often only been able to think within the horizon of the human, especially within Enlightenment thought traditions. As a result, anthropocentrist supremacies are deeply imbued within categories of thinking that we well-intentioned humans use to attempt to extract ourselves from these same anthro-oriented earthviews. A key supremacist illusion, for example, is that only humans create and experience art. How can we unpack this assumption and engage the implications of a more expansive perspective on the matter in theatre practice and pedagogy?

We can be steered away from an indulgent anti-humanist perspective (simply a narcissism in reverse) by remembering, as the authors of a discussion around French thinkers Deleuze and Guattari and the Anthropocene suggest, that "a critique of human exceptionalism and humanist essentialism does not necessarily mean dissolving the specificity of the human into a free-flowing, all-encompassing and chaoid Life" (Saldhana and Stark, 434). Instead, they affirm, it is more realistic to "agree with the humanist tradition that it is incumbent on thought to examine reality *from within* the intractabilities and ambiguities of the human perspective" (ibid., author's emphasis). This involves the productive work of continuously attempting to think beyond the human, beyond the restrictive recurring tropes of traditional humanist epistemologies. Acknowledge where you're anchored, but don't think you're the anchor. A productive posthumanities that will allow us to think and pursue the actor as geoartist, "requires a recognition not only of the politics of the human that informs and produces knowledge of the human; but also of the ontological forces of the nonhuman that press the human from both within and outside" (431). It is Deleuze and Guattari's emphasis on ontology, on the generative and indeed *creative* nature of being/becoming, not reduceable to strictly configured identities as starting points for thought, which makes them particularly well-suited to this discussion.

What then are traditional humanist orientations that can be thought and practised beyond with regard to acting training? The gravitational orbit of identitarian thinking that privileges bounded individuals and objects over processes and events, the notion that subjectivity and agency are solely to be found in the realm of the human and the general acceptance that artistic and creative experience happens in and about human cultures: these and related assumptions burden attempts to create and connect beyond the human. Each of these three errors in thinking has to be challenged, and specific practices need to be put in place in studio training and creation environments to respond to these same challenges, if we are to hope for a more genuine geoartistry in our theatre acting cultures. A fuller account of the actor as geoartist would provide a more comprehensive exploration of the complex intersecting chronologies of how such posthuman conceptualizations have always already been part of performance cultures. I will gesture towards a very limited bandwidth of these generative histories in my account here, primarily elements drawn from canonical Western actor trainers that demonstrate the inherence of posthuman possibility within their work, possibilities that need to be teased out further in order to unpack notions of the centrality of the *anthropos*, if not the identitarian.

Identitarian thinking

Identitarian rather than processual and differential forms of thought privilege theatre training practices that result in centralized and fixed human identities, incarnating implicit dominance over other phenomena. Differential and processual forms of thinking phenomena can generate and lead to specific and singular forms of human expression that acknowledge and celebrate their other-than-human influences and connections.

Contemporary Euro-American actor training, despite various occasional movements and attempts in other directions, is indelibly yoked to a tradition of playwriting (and by extension filmic narrative practices) that prepares actors for the careful expression of complex arcs of human character development through rising and resolving conflict and action. These dramatic homeopathics—the induction and making visible of wide-ranging symptoms of socially and species-anchored tensions for the simultaneous purposes of their intensification and release—traditionally celebrates human exceptionalism and complexity. The other-than-human, be it in the form of "natural," "environmental" or cosmic forces, frequently serves as metaphorical backdrop (think of the role of nature in *Sturm und Drang*), emblematic presence (Chekov's eponymous orchard), or inscrutable divine antagonists (the "fates" from Attic tragedy forward). Relegated to such supporting roles in the narratives central to much industrial and bourgeois theatrical production, the "environment" in its many forms (weather, animals, minerals, technology, etc.), frequently does not feature significantly in actor training save as an opportunity for mimetic emulation by emerging actors as a means of expanding the range of their expression. While exploring and expressing the other-than-human in the context of human ("my character has the soul of a wolf," "my character is anchored like a mountain," etc.) goes some way to having the human recognize its other-than-human resonances and potentials, the general recuperation of such explorations for the purposes of amplifying the register of the human suggests the humanist limitations of such an approach.

If we imagine the body differently, we can think of ways of opening up the *disponibilité* (availability) of the actor's body to other bodies in ways not restricted by limited perspectives that feature the primacy of identity. Maintaining some of the directions offered by Deleuze and Guattari, particularly their taking up of the implications of Spinoza's perspective on bodies, provides useful ways forward in conceptualizing bodies and actor training practices that can follow from this. Rather than the bound, identitarian body of the liberal humanist subject, Deleuze affirms that for Spinoza, a body of any kind, human or otherwise, is constituted by 1) a kinetic component, or an "infinite number of particles" constituted by "relations of motion and rest, of speeds and slownesses between particles"; and, 2) a dynamic component, by which "a body affects other bodies, or is affected by other bodies" (1988, 123). Always already differential in their composition, bodies are expressions of the onto-genetic dynamics at the heart of Spinoza/Deleuze's perspective on ontology; that being is continuously productive and generative, driven by a motor of perpetual dynamic differentiation. Like any other bodies, actor bodies are therefore inherently open and processual, available to be internally dynamized and externally connected with relational intimacies that necessarily exceed the notion of a core discernible character essence embraced by dramaturgies of normative psychological realism.

It is important to recognize that existing canonical actor trainers have a range of tendencies in the direction of differential and compositional understandings of bodies. I'd like to spend a moment looking at Stanislavski and Laban for example, who, although they always recuperate identitarian and transcendent forms of human subjectivity, flirt sufficiently with differential perspectives on bodies that their work can be *détourné* (or "turned," via the Situationist procedure of *détournement*) for postidentitarian and postanthropocentric actor training contributory to geoartistry for the actor.

With his pursuit of *perezhivani*—or "experiencing"— as a key strategy for escaping the declamatory acting and singularizing star-system he resisted,

Stanislavski affirmed the possibility for an actor to creatively engage their organism on a moment-by-moment basis with the flux of events and moments they were always already embedded in. By contrast, Stanislavski asserted, what he calls the "representational" actor can only pursue a selective and reduced engagement with what Deleuze and Guattari would describe as a multiplicity of affects, and even then would only do so in rehearsal circumstances with a view to selecting and curating a small number of such affects for the purposes of *reproducing* them in performance circumstances. Representational actors understand themselves as engaged in reproducing specific moments, whereas experiencing actors seek continuously to initiate processual encounters with affects in such a fashion that simultaneously repeated yet differential moments can predictably, but not reproductively, occur. Representation versus process. We hear curious echoes of Spinoza and Deleuze & Guattari's perspectives on bodies in Stanislavski's desire to have the experiencing actor resonate with differing velocities. Stanislavski writes that as part of their character work actors "make combinations with all sorts of different speeds and measures." Indeed, he suggests that "you cannot get along with just one tempo-rhythm. You must combine several of them" (1989). These combinations can lead to surprise outcomes in which "the overall tempo-rhythm of a dramatic production usually creates itself accidentally, of its own accord" (ibid). The actor, the character, the production: an intersecting, overlapping sequence of tempo-rhythms. And yet, all of this discernment begins and ends with, and ultimately serves, the identitarian construct of the actor, the character, the production. Hints of a vision of compositional assemblages that could be geared towards making fresh connections with any number of other (other-than-human) bodies are foreclosed by the identitarian imagination that always comes back to closed liberal bourgeois subjectivity and the canonical plays that it needs to use to tell its story.

Similarly, Rudolf Laban's work on dance notation, carried over into education, industrial efficiency and theatre training, compartmentalizes movement into various "motion factors" (Weight, Space, Time and Flow) and effort actions (eight combinations of the latter motion factors) with a view to providing the actor and audience the "*sheer experience*" (Gordon, 182) of movement as experienced by actor and spectator. While it is acknowledged that Laban's system is more flexible and refined than the similar systems of Dalcroze and of Meyerhold that preceded his, Laban and many of those who pursue his methods for purposes of acting pedagogy make fundamental distinctions between the kinesphere of the individual and that of the general environment. That is to say, even though Weight, Space, Time and Flow, and the various effort actions are constitutive of both humans and other bodies, for Laban there is a primordial centrality and emphasis on the human individual. This may appear to be a subtle distinction, but for the training of actors and the conception of the art of what constitutes acting that follows from this training, the implications can be significant. The resultant challenge is the following: if the individual human is always already the starting point and the end point in the training for Stanislavski and Laban, how can we imagine using their methods for creating increased connections, sensitivities and characterizations involving other-than-human bodies? I'd like to consider more key theoretical perspectives before returning to representative suggestions on how to organize this *détournement* in teaching/learning studio settings.

Subjectivity, agency and artistry beyond the human

Humans do not have a monopoly on agency. If we ourselves are differential and processual entities like other phenomena, then autonomies of agency and expression are to be found across many phenomena.

A central organizing and unconscious perspective that directs acting trainers (and other humans) to assume the centrality of the *anthropos* is that agency, willfulness and interiority are primarily the province of the human. Central to being able to even conceive of the art created and experience by other-than-human entities that I will discuss below are onto-genetic and self-producing systems (rather than individuals). These depend on a distributed "autonomy of expression" (Deleuze and Guattari 1987, 317) not always already absorbed into and originating from identitarian formations issuing from transcendent evolutive lineages. In other words, there is something that precedes the bound individual body that has a capacity for agency and expression, and this aspect and capacity extends beyond and after that individual subsides. The Deleuzo-Guattarian body of Spinozan influence is constituted of swarms of distributed agency, be it the human body, the tree body, the forest body, the car body, the world body. These swarms of agency that precede and ultimately exceed and escape all identitarian formations (even as they constitute those temporary and metastable structurations that we call identities) generate the ability for thought and creative activity to be expressed.

How can this notion of distributed agency be useful for productively decentring, without necessarily minimizing, human artistry? How can human artistry be viewed in the context of a broader geoartistry? In their final collaborative text, *What Is Philosophy*, Deleuze and Guattari articulate the territories of the three central human pursuits of science, art and philosophy, and argue for the existence of a *geophilosophy* that ties the activity of thought in various ways into the wider processes of territorialization of the earth (85–113). Recognized as a human activity drawing from and gathering bundles of preindividual processes, this radically constructivist (or "inventivist" as Brian Massumi describes it; 2012) and expanded understanding of philosophy situates speculative thought in much broader cosmic and ontological dynamics. As such, geophilosophy speaks to a mode of thinking that both emulates and affirms the multiplicitous complexity of becomings of "natural" reality. This insight is borne from each thinker's commitment to postidentitarian, differential and processual forms of thinking that seek to work outside inherited models of thought anchored in binarist conceptions of substance, ones bound by restrictive logics of recognition and representation (Deleuze 1994, 2004; Guattari 1995). In such a vision, philosophy's role is to generate new concepts, to "bring forth events" and to "[lay] out a plane of immanence that, through the action of conceptual personae, takes events to infinity" (Deleuze and Guattari 1994, 197).

Similarly, they suggest that art, in a world of multiplicity and becoming, is marked by a similar aspiration "to create the finite that restores the infinite: it lays out a composition that, in turn, through the action of aesthetic figures, bears monuments or composite sensations" (ibid.). Art can do this, they argue, because the artist extracts percepts—the essences of perceptions—from the quotidian perceptions they are always already bathed in. Artists also simultaneously extract affects from experienced affections—those shifts initiated by the engagement with percepts—and bundle affects and percepts together within the materials and forms in which they then express themselves: words and paper, clay and figures, bodies and performances and so forth. These "monuments," and art's drive "to raise lived perceptions to the percept and live affections to the affect" (170), take place, as Deleuze and Guattari suggest is the case with geophilosophy, "in the relationship of territory and the earth" (85) and, like this geophilosophy, art can be understood to link "the cry of humanity and the earth's song" (176). The actor-in-training encounters a tree outside the studio and (cue all dismissive clichés about actor training) is asked to "emulate," "explore" or "become" the tree. Attentiveness to the speeds, slownesses and spatialities of the tree and how they resonate in the actor's body, perhaps using language and training from Laban and Stanislavski, provides the opportunity to issue in response a kind of psycho-corporeal artistic expression as part of an ongoing "creative conversation" with the tree. The actor's bundles, affects and percepts in their creative response are the monumental quality of their work of expressing the tree.

Although Deleuze and Guattari are describing human artistic activities, they provide instances elsewhere in their work where they discuss how other-than-human animals engage in artistic activity. In their discussion "1837: On the Refrain" in *A Thousand Plateaus*, they speak of how certain birds go beyond simply marking territory with sound, graduating from "signature" to "style" (1987, 318). This moment takes place when the "expressive qualities entertain variable or constant relations with one another (this is what matters of expression *do*)" (ibid. authors' emphasis). In other words, the expressive qualities "no longer constitute placards that mark a territory but [rather instead] motifs and counterpoints that express the relation of the territory to interior impulses or exterior circumstances" (ibid.). According to Deleuze and Guattari's onto-genetic model, it is clear that all phenomena are creative, and as such engage with a kind of artistry that produces the *art brut* (untrained or "outsider art") of territorial expression. However, a distinction can be made: Namely, they suggest that "what objectively distinguishes a musician bird from a non-musician bird is precisely this aptitude for motifs and counter- points" (ibid.). These qualities, regardless of whether they are "variable, or even when they are constant" serve to "make matters of expression something other than a poster": with these new variations they become a *style* "since they articulate rhythm and harmonize melody" (ibid.).

Central then to the postanthropocentric understanding of art is to be able, at the very least, to speculate about and imagine the experience of the other-than-human entity's relationship to what might be constituted as "monumental" or artistic from their perspective. Indeed, it follows from the discussion of style, although Deleuze and Guattari take their discussion here that birds and other phenomena that graduate from signature to style, from assemblage to monument, may also have a capacity to perceive these stylized, contrapuntal and other-than-human forms of artistry. To think otherwise would be to imply, among other things, that other-than-human moments of artistry are only to be observed and engaged with as artistry by humans, once again engaging in the anthropocentric certainty of the fundamental singularity of the anthropos to be affected by artistic aspects of the aesthetic. I suggest that such instances, both of production and of reception of art by humans and other-than-humans alike, can be described, as with Deleuze and Guattari's philosophy's turn earthwards, as *geoartistry*, or as being *geoartistic*. Both a concept and an unlimited series of artistic practices—phenomena that instantiate events in thought and precipitate the finite that restores the infinite through specific moments of creation—*geoartistry*, like philosophy, can help serve to "*summon [...] forth a new earth*" (Deleuze and Guattari 1994, 99). How do we make sense of geoartistry? To consider the bird's perspective is obviously an exercise in pure speculation, or is it? Are there affective and performative means by which such perspectives—those of the artistic reception of other-than-human phenomena—can be explored in the studio and in training? Can speculation and practice combine to make a potent means of investigating complex aesthetic relationalities between humans and other phenomena?

Implications for actor training of art experienced by other-than-humans and shared between humans and other-than-humans

If other-than-human entities have capacities not only to express but also receive artistic and aesthetic experiences, then actor training practices can draw on the artistic expression of other-than-human entities. Additionally, and more radically, these performances also need to also be geared to other-than-human audiences.

Let's step back for a moment and introduce discussion of becoming before coming back to the question of the artistic and becoming. Two related terms that circulated widely in the philosophical circles from which this chapter draws are the notions of "becoming-animal" and "becoming-imperceptible." The substance of the terms is in some sense self-explanatory, suggesting an emphasis on some form of transformation towards an animality that is other-than-human. Although this is a useful starting point, Deleuze and Guattari clarify that becoming does not constitute "a correspondence between relations. But neither is it a resemblance, an imitation, or, at the limit, an identification" 1987, 237). Instead, a becoming "concerns alliance" (238), alliances between the tempos, speeds and slownesses that comprise a particular body. The actor Robert De Niro, they note,

> "walks 'like' a crab in a certain film sequence; but, he says, it is not a question of his imitating a crab; it is a question of making something that has to do with the crab enter into composition with the image, with the speed of the image".
>
> (*ibid.* 274)

Rather than approaching the animal as something to be engaged with in its totality, becoming-animal invites a consideration of the engagement with *elements* of a particular "other" body in its processual set of complex unfoldings. Deleuze and Guattari specify that becoming-animal occurs "only if, by whatever means or elements, you emit corpuscles that enter the relation of movement and rest of the animal particles, or what amounts to the same thing, that enter the zone of proximity of the animal molecule" (274–275). These molecules are not necessarily little "bits" of the animal, but rather minor, or component aspects available for transformation, that are engaged with. These molecular or "minor" elements are set in distinction from the animal's molar aspects: those simple reduced placeholder sign of the animal's overall identity. They affirm that "[w]hat we term a molar entity is, for example, the woman as defined by her form, endowed with organs and functions and assigned as a subject" (275). De Niro's skill is to engage the complex processuality of his embodiment with the complex processual embodiment of the crab. He is not simply imitating the crab. Thus, in this moment, at least as De Niro and Deleuze and Guattari perceive it, De Niro is becoming-animal, becoming crab:

> You become animal only molecularly. You do not become a barking molar dog, but by barking, if it is done with enough feeling, with enough necessity and composition, you emit a molecular dog. Man does not become wolf, or vampire, as if he changed molar species; the vampire and werewolf are becomings of man, in other words, proximities between molecules in composition, relations of movement and rest, speed and slowness between emitted particles. Of course there are werewolves and vampires, we say this with all our heart; but do not look for a resemblance or analogy to the animal, for this is becoming-animal in action, the production of the molecular animal (whereas the "real" animal is trapped in its molar form and subjectivity).
>
> (274)

Becoming-imperceptible is a more extensive version of the intensivities released via a process of becoming animal. "The imperceptible," they suggest, "is the immanent end of becoming, its cosmic formula," and "becoming-imperceptible means many things" (279). At a basic level, becoming-imperceptible is about rendering one's self open to the compositions of the other bodies making up the world, a "becoming-everybody/ everything, making the world a becoming, is to world, to make a world or worlds, in other words, to find one's proximities and zones of indiscernibility" (280). This radical act of curiosity, openness to an auto-expropriation of rigid boundaries of inside/outside, is "[t]o be present at the dawn of the world" (280):

> One is then like grass: one has made the world, everybody/everything, into a becoming, because one has made a necessarily communicating world, because one has suppressed in oneself everything that prevents us from slipping between things and growing in the midst of things.
>
> (280)

Aren't there becomings-animal and becomings-imperceptible that have little to do with artistry, or appear to at least, such as, for example, when a hunter engages in becoming-animal to be less detectable by his/her prey? Or when, in the well-known example in *A Thousand Plateaus,* when the wasp and the orchid engage in a mutual becoming in order to further their capacities for survival and fullness of expression (10)? The distinction is this: when becomings-animal are undertaken with a view to raising the cry of humanity to the earth's song, when bundles of percept and affect are gathered and the monumental is generated, then the work of human geoartistry is being undertaken. The creativity of onto-genesis is doubled over with an element of reflexivity and art results. Similarly, when becomings-animal are undertaken by other-than-human creatures with the result of some level of expression that can be understood to be artistic, then geoartistry can be also said to be occurring.

I will return to these questions of the other-than-human geoartistic expression in a moment, but first let's turn back to the examples of Stanislavski and Laban and infuse their engagement of tempo, rhythm and spatiality with becoming-animal, becoming-imperceptible. In the spirit of the style of Stanisilavski's *An Actor Prepares*, in which a fictional narrative account is provided to give a sense of how a studio training session is run, we might provide an aspiring actor the following text as a model:

> Thank you for your audition, Mr. De Niro. We are going to give you the part. Are you interested? Okay, please step back into the studio. Thank you. Let me start by stating that your animal imitations are quite thorough, amusing at times even, especially the brachiopods. That said, we get the distinct feeling that your capacity for engaging the crab is superficial, based on the cliché of the "pincer" movement

of the claw, a sideways movement (crabs often move front and back also), and a clicking sound we're not sure that crabs make. In fact, we're quite sure crabs don't make that sound. The crab and sea anemone sequences for the production are going to be central to the overall meaning of what we're trying to get at with this event. There will be cameras in the tidal pool projecting live images of their feeding and mating, their interactions with the sea urchins, starfish, and any small fish that end up in the pool at low tide. Of course, they'll also be interacting with your image, broadcast live on small screens as you respond to them. It might be hard to describe this as a "dialogue," and they'll not necessarily understand the text you'll be working off (the mashup of the Eliot "Prufrock" poem and the myth about the crab and the sea anemone from Senegal). Still we're hoping that all audiences will be affectively engaged by the intensive echoes and synergies of all the complex relationships developed between all performers. We'll start rehearsals and training next with an acting intensive using Stanislavski's tempo-rhythms and Laban's effort actions. This work will be geared towards you understanding your own tempo tendencies, the way you are comprised of space-time-weight components that precede you and extend beyond you. We will then explore through language and movement the ways in which the crab and sea anemone are also similarly constituted, and then begin to explore resonances, echoes, counterpoint, and other relationships between "you," "the crab" and the "sea anemone" that actually foreground how you are each comprised of complex relationalities within your organism, and complex relationships between each of your organisms.

It's now necessary, in response to this sequence on the imagined crab-De-Niro-tidal-pool-sea-anemone performance, to undertake the delicate task of imagining what percepts and affects other-than-human and other-than-animal entities more widely might extract from their experience of the monumentality of the work of these artists. How might the sea anemone, for example, experience its interactions with the intra-assemblage of its own self (territorialized intra-assemblage), and the wider and multiple inter-assemblages of which it is part? Unlike some of the birds that Deleuze and Guattari discuss, whose signal sound becomes style, it might be more difficult to argue that from the human's perspective at least, the progression from signature to style isn't achieved with the sea anemone, given the relative lack of complexity and reflexivity in the sea anemone's actions as compared to the bird's. Birds move quickly, have more complex neuronal pathways—and thus are more liable to have the capacities for articulating creativities beyond the *art brut* phase of expressive territoriality. It might seem to follow that it is more likely that the intra-assemblages of birds can harness the autonomous capacity for expression so as to be able to express style than sea anemones or crabs can. But is not this privileging of individuated expression, with all of its anthropocentric echoes of human agency and autonomy, particularly limited from a postanthro-pocentric perspective that foregrounds *distributed* expression? Indeed, would it not follow that, from the perspective of distributed expressivities, geoartistry in its more complex manifestations of *style* is liable to occur in the sea anemone's relationship with water, light, rock, wind, tide, geo-logical, macro-meteorological and other factors, and over a much more extended period of time? Following this logic, we might assert that in less "complex" organisms, autonomy of expression generates geoartistry in the accumulated complexity of the inter-assemblage over months, seasons and millennia, rather than in the individuated "behaviour" of the bird species privileged by Deleuze and Guattari; one that is more intelligible within an anthropological ethological register. The body of the intertidal zone, the coastline, the hemisphere, the earth, serves as the context for the con-catenation of signatures from a range of smaller bodies contained within these larger inter-assemblages. Dynamic counterpoint—a central element of Deleuze and Guattari's understanding of the progression from signature to style—between the crab, anemone, clouds generated, contributes to the particular distributed expression of the ecosystem of the mountainside. As additional macro factors such as tectonic plates, volcanic eruptions, carbon emissions intervene in the amount of light and precipitation on the moun-tainside, the anemone responds by improvising over multiple iterations and generations of "itselves" to find distributed and different expression in order to move through and meet the trauma and challenge of its own tribulations in a changing world.

As I'm exploring, a significant result of the Spinoza-Deleuze perspective on corporealities is that "a body affects other bodies, or is affected by other bodies" (Deleuze 1988, 123.). Significant for the implications of geoartistry, this constitution and co-constitution of bodies via singular composition as well as mutual engagement with other bodies invites a recognition of the situ-ated nature of ethical engagement. Instead of depending on an arbiter of jus-tice external to bodies to legislate what is or is not moral behaviour, Spinoza suggests that that which is ethical is that which expands a body's capacity to act and be joyful; that which restricts these capacities is unethical. In his reading of Spinoza, Deleuze directs the reader to understand the extra-individual nature of "affections" constitutive of human or any other perceptual experience, in that affections "involve both the nature of the affected body and that of the affecting external body" (*Ibid.* 49). It is then the transmissional and shared aspect of the affections that makes them constitutive of a variety of bodies simultaneously, and that therefore cannot be reduced to the property of one specific body but instead circulate between and among bodies. This circula-tion of constituting and constitutive affections cause "transitions" and "pas-sages" to be experienced between varying states—durational feelings called affects—that allow bodies "to pass to a greater or lesser perfection" (48). This increase or decrease of perfection is oriented not towards a perfection exte-rior to the body with which the body is being compared or adjudicated, but instead towards the functionality and expression of the coherences inhering within and constituting the body itself. Deleuze describes how for Spinoza productive encounters between bodies that mutually enhance their singular velocities and coherences generate shared or joyful encounters: "when we encounter a body that agrees with ours, has the effect of affecting us with joy, this joy (increase of our power of acting) induces us to form the common notion of these two bodies" (118–119).

Given the *creativity* involved in the work of curating such joyful qualities, Deleuze asserts, following Spinoza, that, "[t]he common notions are an Art, the art of the *Ethics* itself: organizing good encounters, composing actual relations, forming powers, experimenting" (119). The creation of the monument of a work of art via the harnessing of affects and percepts constitutes an extra folding of the joyfulness of the expansiveness of bodies, an ethico-aesthetic intensity as Guattari would describe it (1995).

Practically speaking...

Training visions

And after all this talk? The studio awaits (if you're fortunate/privileged), the actor-in-training awaits (if you're fortunate/privileged), the "great outdoors" awaits (if you're fortunate/privileged). After all this conceptual work, which I feel is necessary if we're going to have a solid inspiration and justification for what we're doing (and if you feel otherwise I entirely accept this, although you probably wouldn't have read this far if the conceptual isn't at least *part* of your bailiwick), it's time to imagine futures of training. Those reading this are qualified and interested, so I'm sure you can fill in any blanks. Instead of detailed descriptions, these are simply some admittedly poetic suggestions of what training could involve.

- The actors learn Laban's weight space, time and flow outdoors, or in part outdoors, or bring materials from "nature" (earth, stone, vegetation, animals, water, etc.) into the training studio.
- Tempo-rhythms are explored and shared with a whole range of different phenomena.
- The actors are trained, with due consideration to questions of protection of personal space and safety, to question where their physical body ends and the world beyond it begins.
- The actors are trained, with due consideration to questions of protection of personal space and safety, to spend a week in a natural location remote from human habitation.
- The actors are trained to investigate clouds, skyscapes and constellations.
- The actors are asked to imagine futures in which there are no indoor theatres, only outdoor spaces for performance. What stories would they tell, and how?
- The actors are encouraged to avoid romanticizing or engaging in nostalgic perspectives on nature by seeking to resonate with various types

of technology: gardening tools, engines, computers. How are these metal-, wood- and silicon-based entities different from or similar to carbon-based entities?

■ Groups of actors are asked to tell the story of the earth from its inception to now.

■ Collective and collaborative performances in which humans and other-than-humans share the bill are imagined and pursued. Some of these performances can be planned to last 10,000 years, 100,000 years or longer.

■ The actors are asked to perform for other-than-human entities, living or non-living. What does this mean? What does it look, feel, sound and smell like?

■ The actors are asked to perform for mixed audiences of human and other-than-human entities, either by being in outdoor locations, or by bringing different materials not normally in the studio into the studio. What does this distributed spectatorship feel like?

■ The actors are asked to explore and discover what might constitute the creative and especially artistic activity of other-than-human entities (living and or not living), to resonate with these; to explore what it would mean to perceive them on their own terms; to explore how their modalities might be pursued by human artists.

■ The actors are asked to be other-than-human entities receiving what to them is creative, aesthetic, artistic.

■ The actors are asked to imagine what it would be like to participate in training other-than-human creatures to intensify their own individual or collective geoartistic potentials, and to thereby intensify the joy of shared geoartistic expression.

■ The actors are asked to explore how anthropocentric and traditionally humanist forms of theatre and performance can be understood, even in their anthropocentrism and traditional humanism, to nonetheless be connected to much broader geoartistic forms of expression.

■ The actors are asked to write manifestos of geoartistry for their own practice.

Geoperformance and the future of acting in theatre and performance

These training suggestions for investigation, exploration, refinement of sensibility and capacity revolve around various clear axes: how is the notion of the artistic experienced and expressed in other-than-human entities and assemblages, and how can the human actor connect with, understand and express these other-than-human realities with which they are very much connected? These axes are not pursued in order to simply imitate the other-than-human for the purposes of amplifying the register of human expressive potential. Instead, engaging in shared performances, distributed spectatorships, storytelling that invites humans to understand tales that are not their own and that are not recuperated to Oedipal and humanist pro-jections onto other-than-human experience (the story of the little bird that overcame all odds, impressed its parents, and saved the colony) can be an integral means of intensifying the ethico-aesthetic joy to be experienced by transversal and non-dominant relations between ranges of entities (human and otherwise). This of course doesn't mean an anti-humanism, as I've suggested at the beginning of these thoughts, tempting though that might be in a world where anthropogenic climate change is currently inducing a major extinction event. Human singularity can be expressed and cele-brated, human capacities for refined and distinct types of love, thought and artistry. We do not need to take the crypto-nihilist perspective of con-temporary schools of materialism that see no distinction between a shoe, a human, an amoeba and another shoe. Why not celebrate singularity? As long as we do the important work of anti-dualism and remember some-thing that traditional anthropocentrism has never known or always chosen to forget: that singularity does not have to mean supremacy.

Works Cited

Deleuze, Gilles. *Difference and Repetition.* New York, Columbia University Press, 1994.

———. *Spinoza: Practical Philosophy.* San Francisco, City Lights Books, 1988.

———. *The Logic of Sense.* London, Continuum, 2004.

Deleuze, Gilles, and Felix Guattari. *A Thousand Plateaus: Capitalism and Schizophrenia.* Minneapolis, University of Minnesota, 1987.

———. *What is Philosophy?* New York, Columbia University Press, 1994.

Fancy, David. "Geoartistry: Invoking the Postanthropocene Via Other-Than-Human Art." In *Interrogating the Anthropocene, Ecology, Aesthetics, Pedagogy, and The Future in Question,* edited by Jan Jagodinski. Palgrave, 2018, pp. 217–236.

Gordon, Robert. *The Purpose of Playing: Modern Action Theories in Perspective.* University of Michigan Press, 2006.

Guattari, Felix. *Chaosmosis. An Ethico-aesthetic Paradigm.* Sydney, Power Publications, 1995.

Massumi, Brian, with A. De Boever, A. Murray, and J. Roffe. "'Technical Mentality' Revisited: Brian Massumi on Gilbert Simondon." In *Gilbert Simondon: Being and Technology,* edited by A. De Boever. Edinburgh University Press, 2012.

Saldhana, A., and H. Stark. "A New Earth: Deleuze and Guattari in the Anthropocene." *Deleuze Studies,* vol. 10, no. 4, 2016, pp. 427–439.

Stanislavski, Konstantin. *Building a Character,* Routledge, 1989.

Part 4

Theatre and performance studies/praxis

11 Drawing what you can't see

Meditations on theatre and derangement

Mary Anderson

I've never been good at drawing. I see. But I can't translate. It's not just that my drawings don't look like anything that I want to look at or be with or think about. It's that they evoke neither the illustrative nor the affective capacity to engage me as a viewer. I see. I see. I see. But I can't translate.

My only success—and this really was a psychological success, if any at all—was found in the contour drawing exercises. In these exercises, you look at an object with your eyes while using your hand-holding-a-pencil, to make marks on a page that correlate with what you see. You keep your gaze fixed on the object. Your hand-pencil makes marks on the paper. But you don't look at the paper. You just keep looking at the object. Full disclosure: even this description I've just offered you isn't wholly accurate, as the Encyclopedia Britannica suggests that the exercise involves drawing a single continuous line to render mass, volume and tactile values. Apparently, even when I'm doing it right, I'm doing it wrong.

Anyway, I say it was a psychological success because these drawing exercises, these figurative gestures, are the only ones I can perform without feeling a pronounced sense of shame. In these exercises, I'm released—and somehow permit the drawing to be released—from the kind of elementary burdens of representation that my limited imagination has cooked up around the whole enterprise. The drawing just *is*: an artifact, a record of a performance, an attempt, an *essai*, an act. I don't fault it for its shortcomings. I don't fault myself for my shortcomings (even as I remain keenly aware of them). In fact, the most attractive aspect of my contour drawings are the gaps between what is (perceived to be) and how that perception has been (mis)translated, (mis)construed or otherwise transfigured.

I wonder if my experience with contour drawing can teach me about how to understand my experience of being in the world generally; my perceptions of knowledge, of art, of thought and even of morality. As a theatre educator, I wonder if it might provide a little gateway into how to understand what I ought to be doing or thinking about within my profession, during this epoch of pronounced crisis. I wonder this, because the strategies I see most frequently employed—or *deployed*—in the service of mobilizing relationships and action around the environment through or as art are narrative strategies. They are strategies centrally concerned with representation: representing or otherwise evoking the crisis; illustrating or otherwise exploring the problems we face. They are strategies designed to implicate us, as individuals, in the crisis, and assign us responsibility to behave in particular ways in response to those problems. This is the fundamental premise of worthy and beautiful art like Meredith Monk's *On Behalf of Nature* and the ongoing Earth Matters on Stage Conference and Ecodrama Playwrights' Festival. This is also the fundamental premise of fascinating explorations of forecasted dystopias, such as Toshi Reagon and Bernice Johnson Reagon's musical adaptation of Octavia Butler's *Parable of the Sower*, and the travelling exhibition *The World to Come: Art in the Age of the Anthropocene*. The art of the anthropocene is here, but our relationships haven't changed.

In 2005 Bill McKibben, a renowned environmental activist, suggested that the lack of civic action on behalf of the environment was due, in part, to people's not *knowing* about environmental problems: "Where are the books? The poems? The plays? The goddamn operas?" ("What the Warming"). For McKibben, if art and literature would engage more frequently, more urgently, with the climate crisis, then, he writes, people would *know about it*, it would register in their "gut," and it would become part of "our culture." The reasonable presupposition here is that art has the power to engage people affectively—that the book, the poem, the play, the opera that McKibben seeks can mobilize *information* about the crisis and translate it into *feelings* about the crisis that will alter people's relationship to the environment, their sense of stewardship and their engagement in activism. Katy Waldman echoes this perception when she explains that a literary rendering of the climate crisis "lifts it out of the realm of intellectual knowing and lodges it deep in readers' bodies" ("How Climate-Change"). For McKibben and Waldman, the missing ingredient in the current cultural relationship to the climate crisis is embodiment. And that is a reasonable assumption.

But, in *The Great Derangement: Climate Change and the Unthinkable*, Amitav Ghosh takes elements of these presuppositions about affective experience and embodiment and bends them in a slightly different direction, inviting us to consider alternative reasons for the disjunction. For Ghosh, widespread public disengagement with the problem of climate change is not simply attributable to a lack of opportunity to register its impact, physically and emotionally. The problem stems from what he refers to as an *imaginative failure*. In the domain of literature, politics and public life, contemporary societies are unable comprehend the scale and violence of climate change. And this inability to comprehend—for Ghosh, a form of derangement—stems from our fundamentally problematic modes of thinking and imagining. So, in one sense, Ghosh's take on the problem corresponds with that of McKibben and Waldman, to the extent that all three identify a gap between information and action/behaviour. Further, Ghosh, like McKibben and Waldman, identifies art and literature as spaces and locations of possibility, wherein alternative narratives might emerge. However, there is a limitation on the potential of art and literature that is implicit within Ghosh's argument. Because the same modes of thinking

and imagining that have, thus far, contributed to our inability to comprehend and imagine, are just as influential within art and literature as in
every other facet of our society. So even works as abstract and ingenious as
Monk's *On Behalf of Nature* and the Reagons' interpretation of Butler's *Parable of the Sower* still fall into narrative traps—even the most conceptually and
aesthetically experimental works ultimately collapse, in degrees, within the
confines of their reception. It is in this way that "the Anthropocene resists
literary fiction" (Ghosh 84), suggesting that newer, hybrid forms of art and
experience will be required in order to both register the times in which we
live as well as help advance the very nature of "reading," and its associated
modes of thinking and being in the world.

In what follows, I entertain a set of possibilities about what the philosophical project of Object-Oriented Ontology might have to offer theatre educators in response to the fundamental problem of derangement
that Ghosh articulates. Along the way, I explore the relationship between
the kind of useful, revelatory failure that I find in the context of my own
(in)ability to draw, the circuitous and sometimes seemingly self-cancelling
properties of my thinking about art, and the question of what the "next
now" of theatre and education might look like. From these observations,
I propose that we may discover an unusual theoretical-performative path
into the question of what theatre education could or should be doing in
the context of the many crises we currently find ourselves living through.

OOO

Object-Oriented Ontology (OOO) is a recent philosophical project, inves-
tigated collectively, but broadly and diffusely among many individuals and
across various disciplinary locations. The work of Graham Harman has
been very influential in the development of the OOO project. It has been
further popularized by Timothy Morton and Jane Bennett, who investigate
the poetics and applications of OOO across a range of practices and social,
political and environmental problems. One of the main premises of OOO
is that conventional, human-centred approaches to thought bear distinct
limitations that can be especially destructive when it comes to addressing
large-scale, collective crises: fires, global warming, pandemics...

When I read and listen to the thinkers who are working within this framework, I hear what I consider to be a set of "beautiful questions." The beautiful, in this sense, is a reference to the open-endedness of the questions.
OOO thinkers often speak in terms of "setting aside" particular assumptions
about how systems operate or how values are established, asking *"what if...?"*
Some of the fundamental premises of these inquiries include an embrace of
the following understandings as points of departure and possibility:

- Human understanding and consciousness are finite.
- Objects exist independently of human perception and cannot be onto-
  logically exhausted by their relations with us or other objects.

- Objects have their own agency and exist on equal footing with one another. This includes relationships between objects and humans as well as interactions among non-humans.
- Humans can never exhaust the surplus reality of things. The same is true even of non-human objects in their relations with each other. There is always a surplus unmastered by all our efforts to grasp their properties. A thing is impenetrable to the human senses and intellect.

Based on these shared approaches to consciousness and understanding, Graham Harman posits that "no one is actually in possession of knowledge or truth" (6). For Timothy Morton, embracing this perspective involves the relinquishing of "core beliefs" that govern human thought and action (Morton qtd. in Blasdel, "A Reckoning"). One core belief that has been central to human thinking about the climate crisis is that we must work to address or otherwise solve the problem before we bring the planet to its ruinous end. Morton suggests that it is precisely this "strongly held belief that the world is about to end 'unless we act now'" that is "paradoxically one of the most powerful factors that inhibit a full engagement with our ecological coexistence here on earth" (7). How could this be? Surely the constant reminders that we are on the verge of a crisis are adequately motivating to promote meaningful social and cultural change. Yet Morton says no. Instead, he argues that this kind of "end of the world" thinking is informed by vapid metaphorical constructs that lock us into a kind of dream state in which we move no closer to non-human realities or significant development of human thought and action. Instead, for a number of rhetorical, conceptual and scientific reasons, Morton proposes that we begin from the perspective that end of the world has already occurred, inviting a radical displacement of our perception of the capacity of human reason and turning towards modes of inquiry that destabilize and reorganize our structures and practices of being in the world.

This, to me, seems an exciting position to take up, as it releases us of the perception that human reason necessarily has the capacity to address or otherwise "solve" any particular problem. But if the end of the world has already occurred, why should we bother to do anything? If the end of the world has already occurred, are we released from any obligation to act? What is our motivation in this new paradigm? My interpretation of Morton is not that he is suggesting *it's the end of the world: full stop*. Rather, *it's the end of the world as we know (knew) it*. An acknowledgement of the frailty of the authoritarian paradigm of human-centred behaviours, motivations and assumptions. An exposure of the limitations of human thought and experience. A revelation of the edges of human perception: the edges beyond which there is an overwhelming void. An ending that suggests a beginning.

So, *what then?*

For Morton and Bennett, the answer to *what then?* is to pay even more attention to objects themselves. Bennett has a lyrical, haunting passage in which she describes paying close attention to the way a rat interacts with a glove and a water bottle cap, to create what she refers to as an assemblage that is *in excess* of the rat, the glove, the cap, as well as her perception of either/both/and:

> Glove, pollen, rat, cap, stick. As I encountered these items, they shim-mied back and forth between debris and thing—between, on the one hand, stuff to ignore, except insofar as it betokened human activity (the workman's efforts, the litterer's toss, the rat-poisoner's success), and, on the other hand, stuff that commanded attention in its own right, as existents in excess of their association with human meanings, habits, or projects. In the second moment, stuff exhibited its thing-power: it issued a call, even if I did not quite understand what it was saying. At the very least, it provoked affects in me: I was repelled by the dead (or was it merely sleeping?) rat and dismayed by the litter, but I also felt something else: a nameless awareness of the impossible sin-gularity of *that* rat, *that* configuration of pollen, *that* otherwise utterly banal, mass-produced plastic water-bottle cap.
>
> (4)

In a similar way, Morton is drawn into rapt attention of a painting as an "agential entity," explaining how the painting acts upon him with its own agency:

> I am gripped immediately in the tractor beam of the painting, which seems to be gazing at me as much as or more than I am looking at it … [A]s I approach it, it seems to surge toward me, locking onto my optic nerve and holding me in its force field.
>
> (69)

As part of the larger project of OOO, Morton's description of his encoun-ter with the painting articulates the way in which, even as art objects exert force and influence over human experience, nonetheless "the external world exists independently of human awareness" (Harman 10). If we begin an examination of art and education from the assumption that no one— not the artists, not the audience, not the teachers, not the students, not the bystanders, not anyone—is, per Harman, *actually in possession of knowledge or truth*, what is our call to action? Jane Bennett suggests that we "devise new procedures, technologies, and regimes of perception that enable us to con-sult non-humans more closely, or to listen and respond more carefully to their outbreaks, objections, testimonies, and propositions" (108). To what extent do artifacts of our art-making experience, as agential entities, have the capacity to identify gaps "between knowledge and reality" (Harman 7)? What can we gain—conceptually, imaginatively or even pragmatically— from making slight adjustments to how we approach our encounters with objects? As artist/educators, what forms and relationships are we seeking that acknowledge the limitations of truth as outlined by Harman, yet are still compelled by the transformative, if enigmatic experiences described by Morton?

This brings me back to my contour drawing anecdote and the potential it may have for understanding the function of an OOO orientation to knowledge and experience in an art education context.

Drawn (Again)

One distinction that remains fascinating for me within Harman's thinking is the difference between reality and truth. Whereas there is a shared understanding, within OOO, that no one is actually in possession of knowledge or truth, Harman suggests that we can claim to have some relationship to reality and that articulations of reality remain an important part of our ongoing human work. I am curious as to how this truth/reality distinction maps onto the domain of my drawing experience. Is my failure to produce a drawing that conventionally (or even impressionistically…) registers *as a drawing* a technical failure or a categorical failure? Due to my technical ignorance (or my resistance to becoming technically competent), I am unable to produce drawings that might be understood by other humans as representative of conventional knowledge or truth *as drawings*. Every conventional attempt I make only confirms the absence of knowledge or truth in my product—hence the shame. Even the contour drawing experience—which can, for some, lead to representations commonly understood as reflecting "knowledge" or "truth"—for me, continues to affirm these elements as absent. Yet, because the focus of the contour drawing experience remains principally on the object, itself, I am somehow released of the burden of my self- or societally or disciplinarily imposed perception of failure. The absence of conventionally rendered technical competence remains in my artifact. But the artifact is less an account, a confirmation of my incompetence. It is principally a record of the distance between the object I observed and the reality of my action.

So, how does paying closer attention to objects give us increased access to reality?

In a human-centred model of experience, we are accustomed to thinking about objects in terms of their form as it relates to function. In his essay "The Thing," Heidegger investigates the limitations of this human-function-centred model by exploring the example of a simple jug, which we might first approach or attempt to understand in terms of its capacity to hold water. We might focus on the sides and bottom of the jug as those formal elements that enable the jug to be a jug—if a jug is an object that holds water. But, Heidegger suggests, if we pursue this functional line of thinking more deeply, we discover that it is not the sides and bottom of the jug that do the holding. It is the emptiness of the jug that does its holding. The sides and bottom, as matter, still *matter*. But their importance now is associated with the extent to which they *create the possibility for the empty space*, which does the holding.

This simple shift does not extricate us from the confines of a human-centred approach to objects. We are still (forever, inextricably) bound inside the presupposition that our primary narrative encounter with the jug relates to its water-holding purpose. However, now we are endowing the manifestation of that purpose to the immateriality of empty space as opposed to the material surfaces of the jug's sides and bottom. We are still with the material: next to it, near it, in proximity to the material. But we are focusing on, attending to, the empty space created by the material. This is one potential pathway into agency. The material creates the empty space. The empty space does the holding. (Note that this is a hybrid concrete + metaphorical + metaphysical way of understanding both the material and the emptiness—it lives both in the concrete as well as in the ether.)

As I work on my contour drawings, I watch, I watch, I watch. I listen to the object. I observe. I mark, I mark, I mark as I watch the object I observe. The object reveals various dimensions of itself to me. I mark, I mark, I mark as I observe. But, whether virtuosic or incompetent, the marks record the empty space—the empty space, full of reality, void of knowledge and truth.

The empty space and its tensions

If one strategy for liberating our encounters with objects from their existing entanglements with ideas of what we presume they have been made to do is to reframe them in terms of their emptiness—their potentiality—then how do we hold that space, hold it open before shutting it down with the violence of articulation, the violence of decision, as Anne Bogart writes? Or the violence of the predetermined, *over*determined rhetorical framework that dominates and closes off so much possibility from human engagement with objects in or as art? Since we are holding the space open with the theatrical thinking of Bogart, perhaps we can turn for a moment to Peter Brook and his *The Empty Space*. For Brook, any empty space becomes a stage if you call it such. And a stage space has two rules: anything *can* happen and something *must* happen. So, let us call the empty space that exists in our encounters with objects a stage. Let us open our empty space up to the possibility that anything can happen. But let us also protect our space from the immediate foreclosure of possibility by keeping at bay those forces which would otherwise predetermine or overdetermine outcomes. Because a predetermined outcome or an overdetermined outcome is where art ends. To return to Brook: the empty space we have just declared a stage is not simply a convenient place for the unfolding of a staged novel or a staged poem or a staged lecture or a staged story or a staged narrative. Trying to recreate the idea of something is what Brook calls, unflatteringly, *deadly theatre*. The efficacy of words or actions that transpire in the stage space is dependent, for Brook, on the tensions they create on that stage within the given circumstances.

How do we understand the staging of tensions in this empty space? Bogart opens her chapter on violence with a description of Robert Wilson in rehearsal. In the midst of ensemble warm-ups and other preparations, Wilson enters:

> He sat down in the middle of the audience risers amidst the bustle and noise and proceeded to gaze intently at the stage. Gradually everyone in the theatre quietened down until the silence was penetrating. After about five excruciating minutes of utter stillness, Wilson stood up, walked towards a chair on the stage and stared at it. After what felt to me an eternity, he reached down, touched the chair and moved it less than an inch. As he stepped back to look at the chair again, I noticed that I was having trouble breathing. The tension in the room was palpable, almost unbearable.
>
> (44)

Bogart shares the scene as an artifact of theatre-making, not of theatre itself. But the parallels are clear and, one would imagine, intentional. She moves on, though, to use the moment as a kind of pedagogical message about the violence of decision as it transpires in the theatre-making context: "To place a chair at a particular angle on the stage destroys every other possible choice, every other option" (45). This decisiveness is cruel, as it has "extinguished the spontaneity of the moment," but, for Bogart, it is necessary and even heroic, because "only when something has been decided can the work really begin," and the work of the actor is to bring "skill and imagination to the art of repetition" (ibid.). If trying to recreate the idea of something, for Brook, is what leads to deadly theatre—theatre devoid of the necessary tension to sustain art, as distinct from reiteration—then it would seem that, for Bogart, it is the principal work of the actor to (re)discover freedom, anew in each performance, within the limitations set by the director and designers.

I am writing all of this out and having a number of thoughts. One thought: everyone reading this paragraph about Brook and Bogart—they already know this. Another thought: I was heading in one direction with the Wilson anecdote and now I'm headed back towards a very familiar path. Third thought: does this mean that theatre in/as education already does the thing that I'm calling for with regard to de-centring humans and embracing the thinking of Object-Oriented Ontology? Has theatre already been doing OOO all along? Is this essay finished?

Let me rewind. I've read back through these last several paragraphs numerous times now. I'm stuck on the Wilson anecdote because, before I continued to write deeper into the explanation of how Bogart understands the function of improvisation vs. repetition, what I really wanted to do was think through that Wilson moment *as* theatre. It's a hard thing to do, because as a reader I'm aware that Bogart shared that memory with us, in full knowledge of what it would do for us as readers: help us visualize and sense the moment that a particular stage tension is created for the first time; how precious and complicated that experience of witnessing was for her. And how it illustrated a kind of core aspect of both the making of art and the incredible challenge that an actor faces in the re-creative task of live performance. I want to travel with her easily back down that narrative terrain because it is so true, and because it evokes so many additional questions and thoughts in excess of what immediately resonates in the story and explanation. It's a song I can and do sing again and again. It's why Bogart is so interesting as a thinker, maker, writer.

Tragically, I must interrupt. Back to the tension of the Wilson moment. And this sentence: "Only when something has been decided can the work really begin" (45). I wonder if something is wrong with me. I understand that the dominant theatrical discourse in the academy aligns with Bogart's narrative—even if some of us would question her suggestion that "improvisation is not yet art" (45). My concern is that I think—not just for the purposes of this essay but sort of generally—I think I might rather watch Wilson make decisions in the manner in which Bogart describes rather than watch another play. I understand that Bogart is suggesting that she seeks to create a theatre in which the tensions at work in the Wilson moment are as alive and real as they were in the anecdote. But I suppose, for me, I am feeling an absence in that space of re-creation. I have affection for it. But, in the context of the crises we are living through, marked by the literal devastation of the environment, accompanied by the human failure to comprehend, to imagine and to respond to this devastation as a form of *derangement*, I am drawn more to a theatre of *what happens before something has been decided* rather than a theatre of *what happens only after something has been decided.*

The excess and the void

I've written out that sentence above and now I'm preoccupied with the problems I'm creating here. The exceptions. The inaccuracies. What kind of theatre am I talking about, anyway? Am I just re-invigorating some silly, bifurcated notion about Brecht vs. Artaud? Am I just calling for a theatre of improvisation? A theatre of failure? To all of that: yes. In the sense that contained within what I'm reaching for, as well as my act of reaching, I'm crashing through all that I mean and don't mean, all that anyone else has ever said and not said. All that I want and don't want. Even though, truth be told, I don't know that there is anything I really *don't want*, theatrically. I'd be happy to watch anything Anne Bogart makes, any day. Listen to anything Anne Bogart has to say, any day. Talk about heroes! She's one of mine. So, what's my issue? What is my damage?

I was trying to explain all of this to my husband and the best description I could come up with was swimming in the ocean. The first and last time I swam in the ocean—really swam out *into* the ocean—I was in San Francisco and it was October. That's important, because the Pacific Ocean is fearsome in a way. And on this grey October day, almost no one was on the beach. I really didn't think about the danger and, other than putting on a wetsuit, I didn't prepare. I just got in and swam west. And I kept swimming west. And I kept swimming west. Until, at some point, I stopped and turned around and continued to swim (or maybe at that point the ocean was swimming *me*), while facing the shore. Swimming out west into the ocean was in no way odd, alienating or even frightening, as it likely should have been. I had no perception of a separation between myself and the sea as a kind of continuous effect of past, presence and future. I just was: seabody. But when I paused and thought to turn back and still swim (or be swum by the sea), while facing the shore, this was when I recognized how far out I was, how odd my land-home appeared from this immersion zone. It wasn't that the shore was odd. It was that I became acutely aware of both the excess and the void in my perception. It was, as Bennett described of her rat-glove-cap experience, a "nameless awareness." To call it something would immediately falsify the experience. It was nameless. But, an awareness, nonetheless. It was in excess of my perceived knowledge of anything in particular; in excess of my perceived knowledge of everything in general. It was a recognition of the surplus reality that any *thing*, and really any *concept* represents: always outside of the human grasp. It is in this way that the edges of human perception (including our perception of our perception itself—the edges are not just, spatially, on the outside of perception; there are interior edges throughout) constitute both an excess and a void simultaneously. The void because we cannot fill it with ideas, information, facts, words with which we typically confirm or authenticate our own consciousness. The excess because within that void is not actually emptiness, but rather the vast agency and reality of the external world that exists independently of us, per Harman.

Here is what happens in my contour drawings; here is what happens in my swimming out into the sea; here is what happens—I'm intuiting—when Anne Bogart watches Robert Wilson gazing at a stage, staring at a chair: we become intimately, namelessly aware of the excess and the void beyond our perception. I believe *this* is what is necessary for theatre education in the era of the climate crisis. I believe this is the response to our derangement.

At the time of writing, the U.S. is experiencing an unprecedented reckoning concerning the ongoing crises that have shaped our lives, our relationships, our practices and our systems of oppression. In addition to the environmental crisis, the pandemic crisis and their imbrication within the genocidal oppression of black communities in this country, theatre education housed within universities is necessarily resuscitating the ghost of its own disciplinary crisis. For decades, we have seen a proliferation of scholarship characterizing theatre as a discipline in crisis, a discipline whose future is in jeopardy and a discipline housed within an ideological crisis[1]. Perhaps the very notion of crisis informs our notion of what our discipline is, but we have not become any more adept at responding to crisis. Anne Berkeley, who has documented trends in training in the university theatre in the U.S. in particular, understands theatre's market-oriented curriculum as a survival strategy for a discipline that has been chronically marginalized and misunderstood in the university context for the last hundred years[2]. Carlson echoes the problem of marginalization due to "the troubled position theatre holds in the American culture and consciousness" (118), but argues that theatre academics have compounded this problem, with an inappropriate emphasis on "professionalization," MFA degree programmes, and buildings. Freeman, writing from the UK perspective, notes that theatre education's emulation of the profession is based on a fundamental misinterpretation of outdated theories of the past. "[W]hat is being emulated is an emphasis on narrative engagement through character and plot, linked to the centrality of the written script and immersion in a post-Stanislavskian sense of behavioral realism" (Freeman 5). This "results in a psychologically impelled performance study that favors back story, intention and the stalwart principles of actors' choices," which is "embedded in a legacy of tradition, characterized by hierarchical, imitative and paternalistic structures" (ibid.). If we are to listen closely to Berkeley and Carlson and Freeman, we may conclude that, in many respects, theatre education has embraced conformity and flouted innovation. So then, is theatre education—marginalized through it may be—complicit in the human derangement?

In a recent article that is part of his climate newsletter in *The New Yorker*, Bill McKibben introduces readers to the Climate Interactive EN-ROADS simulator, which

> allows you to change different variables to see what it would take to reduce greenhouse-gas emissions enough to get us off our current impossible track (screeching toward a world something like four degrees Celsius hotter) and onto the merely miserable heading of 1.5 to two degrees Celsius envisioned in the Paris climate accords.

I was initially excited to play with the simulator. But I quickly became disengaged. While the simulator is effective in illustrating that the problem of global warming is "hardwired into our systems, and not solely a function of our habits and choices" (ibid.), the experience of clicking various buttons and shifting various dials so that the temperature goes up or down within a particular period of time is demoralizing. Because, at the moment, it's painfully clear that I have virtually zero control over whether another pipeline is built, whether the governments will roll back another regulation. While my individual choices still matter, McKibben and the simulator make it clear that the climate crisis stems from a massive set of inter-related systemic problems; problems stemming from the derangement; problems that help perpetuate the derangement.

I think theatre education's answer to the derangement of our field and the derangement of our time—of the human condition—is to stage it. In one sense, this might look something like Caryl Churchill's *Far Away*, which exposes, with the playwright's characteristic brilliance, the perversity of human thought in action. But I fear even that simply presenting this work or any other work might prove to only take us down a well-worn narrative path. Recall that Ghosh explains that "the Anthropocene resists literary fiction" (84). The problem, in this case, is not the vision or the execution of Churchill's writing, but the unavoidable collapse of the reception of the work into absurdism, or surrealism. What mechanisms for theatrical engagement exist—or can be created—that simultaneously produce and unravel, that simultaneously weave and tangle, that disturb, that unsettle, that unhinge our perception of our perception? By *mechanisms for theatrical engagement*, I'm not entirely sure what I mean. Frames of reference? Modes of inquiry? Those phrases sound so structurally conventional, I daresay I fear the words, themselves, would imprison their potential.

In his many descriptions of how thought transpires—and the limitations of our understanding of thought—Foucault does not have many "should" statements. But here is one that seems pertinent for the point I'm wanting to make:

> What is essential is that thought, both for itself and in the density of its workings, should be both knowledge and a modification of what it knows, reflection and a transformation of the mode of being of that on which it reflects.
>
> (327)

Up until this particular passage, Foucault has already illustrated how thought already is just that—human thought is perpetually undoing itself, revealing its falsehoods and limitations, its inversions and gaps. But what I hear in this particular "should" statement is something further. The "should" here I take to refer to our understanding of what thought is and how it operates. It's not simply that human thought is replete with fallacies; it's that it is our obligation to attend to those gaps and not simply try to fill them with the right or correct (new) thoughts. It is our obligation to always engage actively in the thought that simultaneously constitutes knowledge and the modification of what is known. It is an embrace of that out to sea experience: Bennett's nameless awareness. It is both a reverence for and a calling into the void.

For Freeman, theatre education's "greatest concern is the crisis wrought by our own misrepresentation of what we do and why" (5). As Berkeley, Carlson and others have suggested, the derangement of theatre education is idiosyncratic to the pressures and complications of the development and sustenance of our field. But it is also paradigmatic of the larger derangement that Ghosh identifies and which Foucault understands as being fundamental to the convolutions of human thinking: "The question is no longer: How can experience of nature give rise to necessary judgements? But rather: How can man think what he does not think?" (323). Foucault's concept of the un-thought offers a fascinating paradox. And as I'm writing out that word, I'm aware that this is the third paradox I've referred to in this chapter. So—rule of three, right?—I must be done? The first paradox was Morton's—when he described the "unless we act now" phenomenon. It's a seeming impossibility—how could the very real *urgency* of action *inhibit* our full engagement with ecological co-existence on earth? It's the structures and conventions of thought—including, but not limited to what we think *thinking* is and what we think thought *does*—that prevent full engagement. The second paradox was the Bogart/Wilson moment, in which I wrestled with the question of how theatre education can (not) do what it already does (not). Because in one sense, theatre education already promotes the kind of attention to the non-human that an object-oriented approach calls for. Theatre education already invites disruption of norms and the unsettling of human assumptions about themselves and their thinking. Yet, somehow, it is so practised in this, so rehearsed in this, that, more often than not, it does *not* do this. And I wonder if one reason for this goes back, in part, to Bogart's observation of Wilson's decision about the chair—the decision that forecloses all other possibilities, the choice that "destroys every other possible choice, every other option." Why do I still gravitate to the idea that the answers to my questions—*how can theatre think what it does (not) think? how can theatre do what it does (not) do?* —may be found in a theatre in which we watch people *make* decisions instead of the theatre that happens only *after* the decisions have been made? But it's got to be more than that. Because what I'm reaching for here—it's not simply a matter of illustrating the extent to which audiences co-produce meaning during the moment of performance and thus, in reality, the decisions are made and made and made. And neither is it simply a matter of advocating exclusively for a theatre of improvisation—which, for the record, I am delighted to watch and, unlike Bogart, think absolutely qualifies as art. I think it's about a theatre in which we watch someone think through something they don't know (yet). They *truly* don't know yet. For Bogart, certain things have to be known in order for the art to happen: the words, the placement of the chair, the footpath of the actor. "Only when something has been decided can the work really begin" (Bogart 45). It is then the work of the actor to "bring skill and imagination to the art of repetition" (ibid). I'm sitting with this. I'm getting it. I understand what she means. But I nonetheless am motivated by a compulsion to ask how the theatre that Bogart is describing—the theatre that we are all familiar with in one way or another—can nonetheless *knowingly inhabit* what *eludes* us—which is, actually, everything.

Practically speaking…

Something interesting to know: derangement as a mathematical concept pre-dates derangement as an etymological moment. Derangement in math involves the reordering of numbers outside of their "natural" state. Instead of [1, 2, 3, 4] in their natural arrangement, you have [2, 1, 4, 3] and [3, 1, 4, 2] and so on. Variations. Like Bach. Some decades after this mathematical moment, around 1737, derangement became officially associated with the "disturbance of regular order." The association of the word with a disturbance of intellect or reason does not arrive until 1800. Here are some questions to consider: what if the derangement that Ghosh describes constitutes both the thought and the un-thought of our potential praxis? We are so accustomed to the idea of the *arrangement*—the scripting and scoring, the composition. What if we were to approach our theatrical experiments with an intimate awareness not only of the paralysis of the intellectual and creative derangement that marks our time but also of the potentiality of derangement as a theatrical tool for radical reordering, intentional displacement, aberrant transposition, bizarre mistake-making and compulsive failure? I am thinking, in particular, of how we can create alternative paradigms for observation, response and development in the rehearsal room that resist or even defy the kind of authoritarian influence of "what works." Even in the most open-ended devised processes, guided by the idea of affinity, "what we love," etc., we are often motivated (consciously and unconsciously) to pursue the development of choices that we think have the most potential to speak fluently within existing theatrical languages. Even when we are seeking something entirely new, we still interpret and pay attention to those devised moments that we think "work" or are otherwise "better choices." When performers create material that we think works less effectively or "doesn't work," that material is either subjected to some form of remediation or, worse, ignored altogether. As an exercise, what might we discover if we approach this work from a space of genuine un-knowing? I believe this might involve setting aside our internal assumptions about what works, what is most effective or has the most potential. In a devising process, what if we were to enter the rehearsal space looking for moments that *don't work*—that appear to be dysfunctional, theatrically, or that otherwise are not "interesting"? What if we then extracted those un-interesting moments and played with them, experimented with them through processes of rearrangement, disturbance, reordering and other tools of derangement and disruption, to see what we could learn from these cumulative failures? One thought is to find a way to withhold or otherwise displace our typical patterns of judgement in response to the material students create. Take, for instance, this description of an invented game from Complicité:

1) Divide your students into groups of about six to eight and give each group a few objects: a rope, a ball, a couple of chairs or waste paper baskets.
2) Ask them to invent a game using the objects they have been given and, as they play it, to refine and re-write the rules.
3) Get the students to present their games to each other.

This is a very exciting open-ended game that many of us might play in our classrooms. The questions that Complicité recommends to follow the exercise are also typical:

■ Which game is most appealing and why?
■ What makes a good game?

- What is the structure of the game?
- Is there a clear end point and a clear winter?
- Do different players have different roles in the game?
- Does the game develop any particular skills?

The withholding or displacement of judgement that one might seek in this exercise might be to structure the invention of the game and the discussion to follow in such a way that students might find the work in the material, rather than through the lens of what is "interesting." How might the material be placed centrally in the development and analysis of the game and its function? One modification might involve asking the students to first create a game based on various materials' intrinsic properties and what you think they do. Balls bounce and are thrown. Chairs are sat upon or used for obstacles or architecture. Waste paper baskets are used as containers to receive the balls. And so on. You could then ask the students to make these materials do something they shouldn't do—or that work against their intrinsic properties. In these exercises, you are pairing yourself with the material as a partner. You haven't eliminated your judgement. But, your judgement is now tempered by this other thing—this material.

Notes

1 See Roach "Reconstructing Theatre/History," 1999; Watson "Actor training in the United States," 2001; Jackson "Professing performance," 2001; Carlson "Inheriting the wind," 2011; Freeman "Drama at a time of crisis," 2012.
2 See Berkley "Myths and metaphors from the mall," 2001; "Changing view of knowledge, "2004; "Theatre in the 'engaged university,'" 2007; "From a formalist to a practical aesthetic," 2011).

Works Cited

Bennett, Jane. *Vibrant Matter: A Political Ecology of Things*. Duke University Press, 2010.

Blasdel, Alex. "'A Reckoning for Our Species': The Philosopher Prophet of the Anthropocene." *The Guardian*, 15 June 2017, https://www.theguardian.com/world/2017/jun/15/timothy-morton-anthropocene-philosopher.

Bogart, Anne. *A Director Prepares: Seven Essays on Art and Theatre*. Routledge, 2001.

Brook, Peter. *The Empty Space: A Book About the Theatre: Deadly, Holy, Rough, Immediate*. Scribner, 1968.

Foucault, Michel. *The Order of Things: An Archaeology of the Human Sciences*. Vintage Books, 1994.

Ghosh, Amitav. *The Great Derangement: Climate Change and the Unthinkable*. University of Chicago Press, 2016.

Harman, Graham. *Object-Oriented Ontology: A New Theory of Everything*. Pelican Books, 2018.

Heidegger, Martin. "The Thing." In *Poetry, Language, Thought*, translated by Albert Hofstader. Harper and Row, 1971.

McKibben, Bill. "What the Warming World Needs Now Is Art, Sweet Art." *Grist*, 22 April 2005, https://grist.org/article/mckibben-imagine/.

———. "What Will It Take to Cool the Planet?" *The New Yorker*, 21 May 2020, https://www.newyorker.com/news/annals-of-a-warming-planet/what-will-it-take-to-cool-the-planet.

Morton, Timothy. *Hyperobjects: Philosophy and Ecology after the End of the World.* University of Minnesota Press. 2013.

Waldman, Katy. "How Climate-Change Fiction, or 'Cli-Fi,' Forces Us to Confront the Incipient Death of the Planet." *New Yorker*, 9 November 2018, https://www.newyorker.com/books/page-turner/how-climate-change-fiction-or-cli-fi-forces-us-to-confront-the-incipient-death-of-the-planet.

12 Coproducing mimesis

Katrina Dunn and Malus fusca

Theatre education, as it is currently conceived, has some big obstacles to face if it wants to make itself relevant to the generation taking part in the worldwide movement of school strikes for climate change. The movement's leader, Greta Thunberg, has deftly called out the obvious irony of pursing a degree in the age of climate apocalypse: "[W]hy should I be studying for a future that soon will be no more" (11)? Moreover, the activist tactic that has come to define that movement and facilitated its mass impact involves a full-on rejection of education. In a present where "children must sacrifice their own education in order to protest against the destruction of their future" (41), schooling has become a site of contestation where the selfish mortgaging of their adult lives by current grown-ups has become the rallying cry for change from a future generation. To position educational environments as places of help rather than harm, theatre programs must also contend with their implication in the complex of anthropogenic forces that have brought about the climate crisis. Una Chaudhuri, early scholar of ecocritical theatre, has detailed "theatre's complicity with the anti-ecological humanist tradition" ("There" 284). Theresa J. May has pointed out how "theatre's artifice has seemed a virtual monument to humanity's triumph over natural forces" (86), and that its exclusionary, anthro-obsessed tradition only "defines drama as conflict between and about human beings" (84). Added to this difficulty is the tendency of many practitioners and theorists of the arts to valorize artistic practice and label it "progressive," regardless of its relationship to the nonhuman. This tendency has met with a stern rebuke recently from writer Amitav Ghosh. Imagining a future perspective, he observes that

> ours was a time when most forms of art and literature were drawn into the modes of concealment that prevented people from recognizing the realities of their plight .... this era, which so congratulates itself on its self-awareness, will come to be known as the time of the Great Derangement.

(11)

Anticipating the angry assessment of future generations, Ghosh predicts that "they will certainly blame the leaders and politicians of this time for their failure to address the climate crisis. But they may well hold artists and writers to be equally culpable" (135). For him, our environmental emergency is also a failing of culture and a shortcoming of imagination; one so serious that it is making our children stay away from school.

Addressing this emergency will challenge theatre and theatre education in many ways, but May and Wendy Arons[1] are correct in identifying a core theoretical task that is central to its success: "theorizing ecological theater and performance will demand a reconceptualization of the nature and purpose of mimesis, and require finding ways to represent the more-than-human world onstage that do not ineradicably 'other' nature" (2). We need a new theory of mimesis rooted in posthuman sensibilities that will reinvigorate our understanding of the nature and function of the stage, the composition and performance of plays, the work of the actor, director and designer, the body of dramatic literature and the study of theatre history. In this chapter we attempt to contribute to that task by touching on familiar theatrical notions of mimesis and blending them with contemporary counterparts from literary ecocriticism, philosophy and science. We propose that mimesis is more appropriately described as an activity that is coproduced, in conjunction with nonhuman kin, and that its uses in a transformed theatrical milieu will lead to new understandings of performativity, storying and ethical representation. May concedes that theatre has also "been a force for activism as well as the dissemination of hegemonic myths" (87). To swing the pendulum in the direction of activist change, and to better speak to our future students, we need new tools. As Elizabeth Johnson suggests, "resurrecting the power of mimesis might be viewed as a way toward an alternative future free from human hubris and ecological catastrophe" (269).

Deep roots

Struggles over mimesis sit at the very core of drama and dramatic theory. In Bert O. States' assessment, the "most important sentence ever written about drama, Aristotle's definition of tragedy as the imitation of an action, contains the whole range of mimetic theory's frustrations and ambiguity" (5). Taking root from seeds planted in Plato's *Republic* and Aristotle's *Poetics,* at "different times and places mimesis has been seen as a reactionary concept, aesthetically and politically, or as a daring revolutionary doctrine" (Gerould 17). For Plato, mimesis was a copy of a copy, an imitation of a lived reality that was itself a degraded version of a realm of pure ideas, and which inspired and rewarded the basest of human emotions. His student Aristotle came to the concept's defence, asserting that the natural impulse to imitate things as they should be (not as they are) could function as a utopian tool for looking towards a changed and improved reality. Although he did not deny the nonhuman the capacity for mimesis, he did use human mimetic virtuosity as a separating device: "man differs from other animals as the most imitative of all" (Aristotle 47). The rediscovery of Aristotle in the fifteenth century reinvigorated interest in mimesis's utopian potential, and Aristotle's version remained dominant until the late eighteenth century when it began to be radically reconfigured. Zola's later formulation of naturalism, as influenced by Darwin, worked to bring theatre into direct contact with a "real life" that was scientifically knowable. Here the goal was no longer utopian but utilitarian and egalitarian, using this new realist naturalism to promote social reform. Co-opted as the predominant aesthetic of the bourgeois establishment, the style soon came under attack from an emerging avant-garde but has continued to hold sway to this day in mainstream dramatic form.

Some critics have suggested that a reason for Plato's animosity towards mimesis is that, when writing developed in ancient Greece, it created a break with oral culture and reshaped the reasoning process. Plato was creating an analytical, "objective" mode of thought based on literacy, as opposed to the more "subjective," participatory oral culture, and mimesis was intrinsic to orality's participatory nature (Nellhaus 23; Havelock 36–49). Likewise, ecocritics have noted the "tendency of alphabetic writing to direct consciousness away from the materiality of the more-than-human world around us towards the ideational world" (Rigby 360).

Conversely, the unsavoury material participation that mimesis encourages has been conceived by some as a kind of contagion (Johnson 275). Religious antitheatrical literature has long maintained that the experience of performance goes beyond representation: "Only the filthiness of plays and spectacles is such as maketh both the actors and beholders guilty alike" (Munday 66). Citing "Plato's contention that imitation is formative and that you risk becoming what you enact" (21), Jonas Barish gives numerous examples in *The Anti-theatrical Prejudice* of a destabilizing theatrical contagion, to be avoided "lest the custom of pleasure should touch us and convert us" (Northbrooke 4). Fear of conversion-by-contagion has fashioned a distaste for theatrical imitation that merged in the early twentieth century with an attack on bourgeois realism as an essentially conservative artistic form, based on soothing familiarity and recognizable stereotypes. Bertolt Brecht disliked the tendency for mimesis to encourage conformity to existing social orders, and instead created a new mid-century version based on demonstration rather than imitation (Gerould 36–37). Samuel Beckett attacked the mimetic basis of the theatre, restricting actors' creative freedom, and fragmenting and decontextualizing stage action. In theory, poststructuralists marginalized the imitation of life, labelling it naïve, reactionary and epistemologically false, and the early twenty-first century gave rise to a crop of new anti-mimetic theatrical forms that were theorized by said poststructuralists as "postdramatic."

Interest in this feud has given rise to some new theories of mimesis that somehow straddle the divide, merging both the conservative and revolutionary potentials in mimesis to form intriguing hybrids. Elin Diamond's *Unmaking Mimesis: Essays on Feminism and Theater* calls for a recapturing of mimesis's playful and destabilizing aspects, and by reading it in conjunction with feminist theory, details the ways mimicry can encourage difference. For her, "mimesis is and has always been more than a morphological issue of likeness between made objects and their real or natural counterparts. Mimesis is also epistemological, a way of knowing and therefore valuing" (104). This notion that mimesis is not a mechanistically numbing copying, but instead a process of generating new knowledge and realities both creative and political, is also to be found in Homi K. Bhabha's exploration of post-colonial mimicry. Bhabha sees mimicry a tool for subtly renegotiating the terms of the colonial context with a "hybrid form of resistance and obedience, taking on the voice of the oppressor but with elements of parody or resistance" (Stevens 199). The colonized can outwit the colonizers without recourse to violence, but with a productive, ironic dissent. What emerges from this mimicry process is a "third space" of mixture, hybridity and difference, highly productive and unique to the society it creates. It is a kind of double vision which, skewering both the familiar original and the newly threatening copy, discloses the contradictions of colonial experience and disrupts its authority (Bhabha 121–131).

Twin trunks

In climate change we see a similar mix of familiarity and menace. Extreme weather as well as our cherished high carbon lifestyles are recognizable, but the tsunami of guilt that results from owning the connection between the two generates what Ghosh has called the "environmental uncanny" (30). Importing Freud's concept of the uncanny into ecological discourse (see also Morton 177–178), he observes that something as mundane as the weather has been rendered mysterious and threatening while also stirring a sense of recognition. In the anxiety of the uncanny we "recognize something we had turned away from … the presence and proximity of non-human interlocutors" (Ghosh 30). The uncanny reminds us of what we have forgotten; that the nonhuman world is not an endless stream of dead resources designated for our comfort, but precious, finite, agentic and communicative. In the dawning horror, when the bubble of a traditional humanist world-view bursts, the well-worn uses of mimesis prove weak tools. Jeffrey Jerome Cohen articulates how early attempts at reading representation from an environmental sensibility are "typically concerned with how nature is represented within a text and how models of human inhabitance unfold within an imagined natural world." This work "often focuses on the destabilizing encroachment of industrialized society into wild spaces, the restorative and even ecstatic powers of unblemished landscapes, and the companionless dignity of nonhuman creatures" (*Prismatic* xx). Cohen also points out that these kinds of readings, while working to use mimesis as a salve for the anguish of the environmental uncanny, "tend to reproduce what Bruno Latour calls the Great Bifurcation, a split between nature and culture" (*Prismatic* xx; Latour, "Attempt" 476–484). They reinforce the isolation of human subjects in their cultural milieu and the unknowability of the nonhuman object in the vacuum of nature, oblivious to the fact that just outside of the edge of awareness both are one. How did our perception of nonhuman kin come to be so resolutely and completely suppressed, and can a reimagined and repurposed engagement with imitation possibly encourage some healing effect?

The search for a new form of posthuman mimesis is proceeding apace, but reaches as far back as the work of Walter Benjamin. In 'The Doctrine of the Similar" and "On the Mimetic Faculty," Benjamin maps out a vision of mimesis unconcerned with the explanations of Plato or Aristotle, instead serving it up as a prelapsarian remedy to the alienation of modernity:

> The gift that we possess of seeing similarity is nothing but a weak rudiment of the formerly powerful compulsion to become similar and also to behave mimetically. And the forgotten faculty of becoming similar extended far beyond the narrow confines of the perceived world in which we are still capable of seeing similarities.
>
> ("Doctrine" 69)

The idea that our current facility for mimesis is but a vestigial remembrance of a formerly powerful and nuanced way of knowing and creating sets up imitation for an explosive rebirth and suggests that theatre is a source of a radically new way of experiencing the world. Benjamin dares to step outside of the human realm when explaining our propensity for mimesis: "The child plays at being not only a shopkeeper or teacher but also a windmill and a train" ("On" 333; qtd. in Diamond 154). We learn by imitating and inhabiting both the animate and the inanimate. Benjamin also labels our mimetic faculty "phylogenetic" in origin ("On" 333; qtd. in Diamond 153–154); part of our biological being and connected to the entire living world, rather than a learned cultural attribute. In this he anticipates a twenty-first century biomimetic revolution in which the sciences are leveraging imitation to rescue our future.

A model of mimesis rooted in the structure of the brain comes from Iain McGilchrist's *The Master and His Emissary: The Divided Brain and the Making of the Western World*. McGilchrist's theory asserts that human beings have two fundamentally opposed capacities rooted in the bi-hemispheric structure of the brain. Though they cooperate, they are also involved in a power struggle that "explains many aspects of contemporary Western culture" (3). The left hemisphere tends to deal with pieces of information in isolation while the right deals with phenomena in their wholeness. He describes mimesis as "the meta-skill that enables all other skills" (253), generating rapid expansion of the brain in early hominids, and leading to current human brain complexity. He spends much of the book attempting to map shifts in the balance of power between the two hemispheres and asserting that western culture has shifted further and further towards the action of a dysfunctional left hemisphere, increasingly mechanistic, fragmented and decontextualized. For McGilchrist, mimesis is the revolutionary force that can release us from the determinism of a world created by hemispheric imbalance. Imitation "introduces variety and uniqueness to the 'copy', which above all remains alive" (247), and allows our brains to escape from the confines of our own limited experience. The activity of mirror neurons in the brain demonstrates that when we imitate something it is as if we are experiencing it, and this experiential, engaged copying is "imagination's most powerful path into whatever is Other than ourselves" (248). Mimesis moves us towards the embodied nature of our existence in its interrelatedness and wholeness, as promoted by the right brain.

Another branch of the same tree, Janine Benyus' *Biomimicry: Innovation Inspired by Nature*, calls for a revision of design, engineering, invention and resource management enabled by mimicking natural production. Observing that we have "practiced a human-centered approach to management, assuming that nature's way of managing had nothing of value to teach us" (3), she insists that the key to our survival lies in imitating nonhuman expertise that has already solved many of the problems with which we are grappling. Like Benjamin and McGilchrist, Benyus insists on the biological roots of human mimesis: "we as a species are well shaped to mimic what we see and hear" (295), but unlike Aristotle she does not see this as a predominantly human purview:

> We are not the only species to have prospered through imitation …. There are behavioral mimics like the cowbird chick, coloration mimics like the viceroy butterfly that resembles the poisonous monarch, and shape and texture mimics like the walkingstick, an insect that looks like a twig. Biomimicry helps animals and plants blend into their surroundings, or, in the case of the viceroy and monarch, to take on the traits of a species that is better adapted to its environment. By imitating nature's best and brightest, we, too, have a chance to blend in and become more like what we admire.
>
> (296)

Benyus' three basic principles of biomimicry—nature as model, nature as measure and nature as mentor—tap into much of the aesthetic, ethical and epistemological struggles that have surfaced in human relations with the nonhuman over millennia. Most of the ethical questions she broaches can be reduced to the simple question: "[I]s there a precedent for this in nature" (291)? Benyus, however, will admit that even a thoughtful version of biomimesis has its risks. "The last really famous biomimetic invention was the airplane (the Wright brothers watched vultures to learn the nuances of drag and lift). We flew like a bird for the first time in 1903, and by 1914, we were dropping bombs from the sky" (Benyus 8; qtd. in Barad, "Queer" 315).

Entangled branches

Philosopher and ecocritic Karen Barad[2] asserts that "biomimesis is a particularly poignant call for the incorporation of difference at every level" ("Queer" 334) and reconstitutes humans in a far broader scope of existence. We (humans) and they (nonhumans) are mutually entangled, and these same entanglements are being constantly undone and reconfigured. Mimesis can help us call into question "the dualisms of object/subject, knower/known, nature/culture, and word/world" (Barad, "Posthuman" 137), and with that questioning the curtain falls on humanity's triumph over natural forces. Dipesh Chakrabarty has shown how "the wall between human and natural history has been breached" by anthropogenic explanations of climate change (349); human drama and the drama of the earth's geologic and biotic unfolding are now one and the same, demoting human neurosis to a mere bit part. Leading ecocritics have worked to label the new texture of our entangled existence: it is for "Latour a network, Morton a mesh, Stacy Alaimo transcorporeality, Tim Ingold a meshwork, Giles Deleuze and Félix Guattari an assemblage" (Cohen, *Prismatic* xxv). We need new methods of configuring imitation in theatre based on this entangled flat ontology that explore "the materiality of the thing formerly known as an object" (Bennett xvi) and recognize the agentic contributions of nonhuman forces.

In exploring our new relational existence we should proceed gingerly, with a focus on making and remembering kin. Donna Haraway's notion of kinship extends subjectivity to nonhumans and looks to explore how all entities are boundaryless and co-constitute each other (*Staying* 99–103). Some of this kinship can be forged through artistic practices, or versions of what Jane Bennett calls "a human-nonhuman working group" (xvii). A core principle of these working groups must be to recognize nonhuman agency—the ability to communicate, duplicate and make change—even in the inanimate. Bruno Latour's actor-network sees all action, including human, as the outcome of networks of association between a host of different actants ("What" 235–236). Until recently we consumed our way through the spoils of an inert earth, content with cataloguing its contents. "The only room such models typically leave for the agency of forests, streams, weather and mountains is a pressing back in the form of cataclysm" (Cohen and Duckert 5). The current cataclysm of the climate has revealed to us the extent to which our actions are interwoven with a host of others, and that, in working with them, we need to proceed democratically and extend the rights that *we* cherish to millions of species and billions of assemblages, thus diminishing the weight of our own influence in the world (Latour, "What" 233–236). As Floyd Favel suggests, in pursuing this course, we are "subjected to the laws that are not made by humans" (122).

A more democratic and perhaps productive notion of mimesis than the ones considered thus far suggests that representation is not something that we do to reality, but rather something that the nonhuman does to us. Practices of making and knowing, including theatrical creation with its mimetic tendencies, cannot be fully claimed as human practices. David Fancy, glossing Gilles Deleuze, recommends "repositioning that which is affected ... to become that which *gives*" (65). A posthuman mimesis will open us up to how our surroundings *think through us*. Historically, the nonhuman has been excluded from the rights and privileges of humans, and that includes artistic creation. As Haraway notes, we've placed them "outside the order of signification, excluded in trading in signs and wonders" (Haraway, *Modest* 8; qtd. in van Dooren and Bird 4). Assuming that the inanimate and the biotic could never participate in such trade commits once again the sure-to-be-disproved error of gifting to humans special faculties that no other earthly inhabitants possess. It also keeps us lonely in our creation process. Opening up to the multiple coproducers directing us and working with us is key. As Rigby says, if "other than human signifying systems are not to be rendered mute and their potentially resistant agency denied" we must acknowledge that our constructions are already constrained "by a more-than-human material reality that precedes and exceeds" them (364). Evidence of this relational coproducing is already with us in the strata of our creative inheritance, though pressed thin between heavy layers of human self-obsession. One task for critics will be to bring this evidence to light, learn to read it and recognize it in the same way we have other marginalized voices. Going forward, we must assume that the copied-but-new entities given to us by nonhumans through the imitation process possess agency which might be resistant, or celebratory, or calculated, or perhaps something we cannot yet imagine.

Pink flowers

The pursuit of posthuman mimesis immediately encounters the dual dangers of anthropomorphism and transcendentalism. Certainly, theatre and performance have long histories of damaging anthropomorphism as part of a hefty artistic tradition that "runs from ancient beast fables to Donald Duck" (Chaudhuri, "(De)facing" 13). These portrayals mimic other life forms for the purpose of discussing human neuroses and conflicts, thus re-crowning the *anthropos* in the hierarchy of life. To counter this tendency, Martin Puchner has introduced the idea of "negative mimesis," a tactic in the sphere of theatre and performance that "demonstrates the extent to which the very distinction between humans and animals is the product of projection and representation" (21). Negative mimesis creates a breakdown of representation in which the nonhuman can negatively appear. Instead of focusing on mimetic portrayals of other life forms, it highlights the mechanisms by which we instrumentalize our kin and learn to unsee our relational ties. Similarly, Timothy Morton has tackled sentimental transcendentalism with his concept of "ecomimesis." For him, ecomimesis "operates whenever writing evokes an environment" (33). It is constructed to melt our hearts and make us love nature, and often materializes in environmental writing in the form of "as-I-write-these-words" reflections describing the weather, flora or fauna around the writer. Janine Benyus has an instance of this in the conclusion to *Biomimcry*: "As I put the finishing touches on this book, two households of geese fuss in the pond right outside my window" (285). Morton labels ecomimesis an error and sees in it a desire to rupture the gap between us and the world with weak aesthetic means; to bring the word somehow closer to the goose. Instead, the attempts are hollow and function to help us sidestep "the obligation to encounter non-identity …. Its mode is one of avoidance," and its effect is similar to that of a laugh track in a sitcom (125). He insists that we need to interrogate the atmosphere that this kind of conjuring creates in order to root out our tendency to fetishize and aestheticize the nonhuman. Failure to do this interrogative work results in an act of bad taste: "In seeking to become non-art, ecomimesis becomes poetical, a kitsch embodiment of the artistic aura itself" (132).

While these dangers have their pitfalls, mapping our interrelatedness in the theatre, while being hypervigilant about bad taste, might severely hamper the exploration. Perhaps there is a productive, hybrid mode of these forms of mimesis such as Diamond and Bhabha have discovered with other forms of mimicry? Some theorists, such as Jane Bennett, would actually argue that these derided forms provide important stepping stones to unlocking nonhuman vibrancy: "We need to cultivate a bit of anthropomorphism – the idea that human agency has some echoes in nonhuman nature – to counter the narcissism of humans in charge of the world" (xvi). In fact, offering our understanding of our own agency and complexity may be the only way to revivify our hardened and lifeless concepts of the nonhuman. As Latour says, "[n]othing is more anthropocentric than the inanimism of nature" ("What" 234). What is certain is that inviting in the nonhuman will wildly disrupt our aesthetics, and we need to be open to what this will afford us. If mimesis is contagion, we will be infected, and we will change. There will be clumsy attempts and embarrassing failures. As novelist Richard Power says of humans and trees, they "will learn to translate between any human language and the language of green things. The translations will be rough at first" (496). In the theatre, to learn new modes of communication, we will have to expose ourselves to contagion, and we will have to risk kitsch.

*As I write these words the tree outside my window is busy dragging water up from the ground, processing light from the sky, and pushing seeds into fruit that will ensure its reproductive dissemination far and wide. A crabapple tree (*Malus fusca*), its affair with a human radically altered its course of life. A determined arbor sculptor bifurcated the trunk and sent its twin halves expanding in different directions, then bent them to grow around each other, like the torsos of two Graham dancers spiraling in opposite directions, or conjoined twins grappling in a fight. This* Malus fusca *is a showstopper. Strolling people halt and stare, motorists get out of their cars and take pictures. In truth, I chose the dwelling for the tree, to have the king's seat for its cycling performance of awe-inspiring transformation. In the winter it is black, hopeless, barren; Beckett's* Godot *tree before its act two leaves appear. Spring heat slowly releases the headdress of green waiting patiently inside. Then pink flowers suddenly appear, covering its entire surface area for a week before blowing away completely in a matter of hours. Then the fruit – so much fruit - green at first and then red, hanging heavy from the branch and then pummeling the ground with its abundance. Birds, squirrels, and rabbits follow the fragrance to the tart repast,*

sometimes staggering away drunk on the fine wine of late-season fermenting crabap-
ples. Enormous deer leap over the fence in the dead of night to feast and carry seeds
away in their digestive tracts. Well done, Malus fusca.

   The tree is getting busy with me too. We're making something together.

Sour fruit

It is theatre's embodied quality, its foundation of participatory oral expression that so repulsed Plato, but that also makes the form so finely attuned
to the revolutionary potential of imitation. As Elin Diamond says, "[t]he
body is mimesis's force field" (153). The actor's body holds the capacity to
enter the meshwork, the assemblage, in an exchange of meaning-making
with nonhuman others. The actor responds to the call, enters the entity
being imitated (or allows themselves to be entered) and pulls the audience
in as well. Time and space are altered, and experience is exchanged in
densely loaded packages of sensations, images and ideas. Actors combine
physical mimesis with their own original imaginative work to generate representations that are new with each iteration, perhaps because the actor
makes a change, or perhaps because their coproducing partners change in
nature, or intention or both. The actor's embodied imitation can change
over time, be contested, overwritten and then rediscovered again. A great
example of what Karen Barad calls "material-discursive phenomena" ("Posthuman" 141), theatrical mimesis is being, thinking and feeling all at once.
As a tool to explore the posthuman world it holds unparalleled promise.

   Redefining performativity, inspiring new ways of creating story and forging an ethical pathway for creators should be the first order of business of a
posthuman theatre. Barad has reworked Judith Butler's idea of performativity for a relational world: "performativity is not understood as iterative
citationality (Butler), but rather as iterative intra-activity" ("Posthuman"
146). We are constantly redefined by our fluid relations with the nonhuman
and the gestures and speech acts that emerge from those mixings. Donna
Haraway has called these relations "sympoiesis" or "making-with" (*Staying*
58–98). Diamond, glossing Butler, asserts: "to mime is to participate in
what is mimed" (173). Moreover, this posthuman performativity values difference in the extreme, generating hybrids from a wide range of mimetic
experience. Finding nonhuman story forms will also define the new theatre practice and embed a coproduced mimesis in unique structures with
unprecedented uses of time and space, and new understandings of what
constitutes an audience. van Dooren and Bird write that story "emerges
out of an ability to engage with happenings in the world as sequential and
meaningful" (3), and thus is an attribute of much of the nonhuman world.
Likewise, Latour says "[s]torytelling is not a property of human language,
but one of the many consequences of being thrown in a world that is, by
itself, fully articulated and active" (Latour, "Agency" 13; qtd. in Menely
and Taylor 2). Nonhuman forms of storying will challenge methods of
theatre-making and art-making, and therefore must be credited as such.

Storms will coproduce visual art, props will cocreate plays, trees will cowrite essays. A properly coproduced mimesis will also help us with ethical limits on our actions and for our art-making. As I note above, Benyus' ethical touchstone for biomimesis is that if it can't be found in nature, there is probably a good reason for its absence. It casts the nonhuman world as mentor, and one that will likely also make ethical claims on us. Barad warns, however, against rooting the new ethics in erroneous notions of nature-as-purity: "there are copious examples of misguided attempts to enlist Nature as justification for every possible social prejudice, including racism, sexism and homophobia" ("Queer" 317). Instead we should look to our mentor's propensity for difference, hybridity and constant transformation.

As we navigate what Kathryn Yusoff calls this "epochal swerve" (2), the theatre will change as it has many times before. How it will change and how we will teach that change will decide whether the generation of school-strikers will find value in learning our embodied, mimetic ways, or whether they will avert their eyes from a Titanic theatre plunging into oblivion, still dancing to the tune of the Great Derangement. Iain McGilchrist agrees that mimesis holds a roadmap for this change: "The great human invention, made possible by imitation, is that we can choose who we become, in a process that can move surprisingly quickly." His enthusiasm continues with a warning: "we had better select good models to imitate, because as a species … we will become what we imitate" (253). By allowing the nonhuman its due as an active participant in our change we reinvigorate mimesis's revolutionary potential and flood the stage with a host of collaborators from which to learn and with whom to interface. Not all of this relationality will feel good. Sometimes it will bear sour fruit, nourishing but hard on the palate. As the climate begins to roil and boil our interrelatedness may even start to feel like a prison with some kin guarding the door and others administering punishment. For now though, we offer a new version of Morton's ecomimesis to fuel our fearless surrender to kitsch. With "apocomimesis," "as-I-write-these-words" reflections will draw the reader or audience member into the calamity outside the doors: the hottest-summer-on-record, or the steadily rising waters. Cue the laugh track.

I write these words in times of contagion. The doldrums of quarantine slowly waddle in turtle time across a vast expanse of solitude. Children are away from school, but not to protest the future we stole from them. They stay home to keep from inhaling each other's droplets. Their educations are jeopardized to stop a state-of-the-art virus from ripping through our bodies like a clear cut, or from riding us like a jet stream swiftly around the world and then back around once more. Well done SARS-CoV-2: We now know that we are all connected.

Practically speaking…

In this chapter we suggest that we might consider all mimesis to be coproduced, and that encouraging students to reconsider the singularity of their own creations could help them to connect more deeply with nonhuman actors. In the studio, this might mean abandoning traditional roles to instead collaborate with plants, animals, a space or place, organic or inorganic objects. There are glimpses of this already in theatrical lore. Actors can speak of how a costume *gave them the character.* In a coproduced mimesis this kind of collaborative facility would be greatly extended, and the pool of potential collaborators significantly widened. In the classroom, works of dramatic literature might be read and researched with a focus on finding and naming nonhuman contributors, and papers might be cowritten much as we have done here. Our suggestion is to begin with crediting: "A play by second-year theatre students and the field behind the Administration Building," or "Written by STUDENT NAME and a metal chair." Preemptively honouring a creative contribution will help to open students to the presence and value of the many nonhuman interlocutors that we have all previously failed to perceive or acknowledge.

Notes

1 Wendy Arons and Theresa J. May are co-editors of *Readings in Ecology and Performance*, one of the first collections of essays on this set of topics, and a source that introduced a number of important questions and areas of investigation. They and their contributors are cited elsewhere in this volume.

2 In *The Performative Power of Vocality*, Virginie Magnat argues compellingly for the ethical obligation that non-Indigenous scholars who engage with posthumanism have an to cite Indigenous scholarship. She details the ties between posthumanist philosophy (especially the work of Karen Barad) and Indigenous philosophy which has a long history of valuing human/nonhuman relationality and nonhuman agency. Respectful anti-colonial engagement with posthumanism must seek out and recognize Indigenous sources (124-186). Though I have cited the work of Indigenous scholar Floyd Favel in this article, I recognize that this is only a beginning and that there is much more work that can and should be done in this vein.

Works Cited

Aristotle. *Poetics*, translated by James Hutton. W. W. Norton & Company, 1982.

Arons, Wendy, and J. Theresa May. "Introduction." In *Readings in Performance and Ecology*, edited by Wendy Arons and Theresa J. May. Palgrave Macmillan, 2012, pp. 1–10.

Barad, Karen. "Posthuman Performativity: Toward an Understanding of How Matter Comes to Matter." In *Material Feminisms*, edited by Stacy Alaimo and Susan J. Hekman. Indiana UP, 2008a, pp. 120–154.

———. "Queer Causation and the Ethics of Mattering." In *Queering the Non/human*, edited by Noreen Giffney and Myra J. Hird. Ashgate, 2008b, pp. 311–338.

Barish, Jonas A. *The Antitheatrical Prejudice.* University of California Press, 1981.

Benjamin, Walter. "Doctrine of the Similar." In *New German Critique*, No. 17, translated by Knut Tarnowski. Spring 1979, pp. 65–69.

———. "On the Mimetic Faculty." In *Reflections: Essays, Aphorisms, Autobiographical Writings*, translated by Edmund Jephcott, edited by Peter Demetz. Schocken Books, 1978, pp. 333–336.

Bennett, Jane. *Vibrant Matter: A political ecology of things.* Duke UP, 2010.

Benyus, Janine M. *Biomimicry: Innovation Inspired by Nature.* Morrow, 1997.

Bhabha, Homi K. "Of Mimicry and Man: The Ambivalence of Colonial Discourse." In *The Location of Culture,* by Homi K. Bhabha, Routledge, 1994.

Chakrabarty, Dipesh. "The Climate of History: Four Theses." In *Ecocriticism: The Essential Reader,* edited by Ken Hiltner, Routledge, 2015, pp. 335–352.

Chaudhuri, Una. "(De)Facing the Animals Zoösis and Performance." *TDR The Drama Review,* vol. 51, no. 1, 2007, pp. 8–20.

———. "'There must be a Lot of Fish in that Lake': Toward an Ecological Theater." *Theater,* vol. 25, no. 1, 1994, pp. 23–31.

Cohen, Jeffrey J. "Introduction: Ecology's Rainbow." *Prismatic Ecology: Ecotheory Beyond Green,* edited by Jeffrey J. Cohen, University of Minnesota Press, 2013, pp. xv–xxxv.

Cohen, Jeffrey J., and Lowell Duckert. "Introduction: Eleven Principles of the Elements." *Elemental Ecocriticism: Thinking with Earth, Air, Water, and Fire,* edited by Jeffrey J. Cohen and Lowell Duckert, University of Minnesota Press, 2015, pp. 1–26.

Diamond, Elin. *Unmaking Mimesis: Essays on Feminism and Theater.* Routledge, 1997.

Fancy, David. "Geoperformativity Immanence, Performance and the Earth." *Performance Research: On Participation,* vol. 16, no. 4, 2011, pp. 62–72.

Floyd Favel. "Theatre: Younger Brother of Tradition." *Indigenous North American Drama: A Multivocal History,* edited by Birgit Däwes, State University of New York Press, 2013, pp. 115–122.

Gerould, Daniel C. "Introduction: The Politics of Theatre Theory." In *Theatre, Theory, Theatre : the Major Critical Texts from Aristotle and Zeami to Soyinka and Havel,* edited by Daniel C. Gerould, Applause Theatre and Cinema Books, 2000, pp. 11–42.

Ghosh, Amitav. *The Great Derangement: Climate Change and the Unthinkable.* University of Chicago Press, 2016.

Haraway, Donna J. *Modest_Witness@Second_Millennium.FemaleMan_Meets_OncoMouse: Feminism and Technoscience.* Routledge, 1997.

———. *Staying with the Trouble: Making Kin in the Chthulucene.* Duke UP, Durham, 2016.

Havelock, Eric A. *Preface to Plato.* Harvard UP, 1963.

Johnson, Elizabeth R. "Reconsidering Mimesis: Freedom and Acquiescence in the Anthropocene." *South Atlantic Quarterly,* vol. 115, no. 2, 2016, pp. 267–289.

Latour, Bruno. "Agency at the Time of the Anthropocene." *New Literary History,* vol. 45, no. 1, 2014, pp. 1–18.

———. "An Attempt at a 'Compositionist Manifesto'." *New Literary History,* vol. 41, no. 3, 2010, pp. 471–490.

———. "What is to be Done? Political Ecology!" In *Ecocriticism: The Essential Reader,* edited by Ken Hiltner, Routledge, 2015, pp. 232–236.

Magnat, Virginie. *The Performative Power of Vocality.* Routledge, 2020.

May, Theresa J. "Greening the Theater: Taking Ecocriticism from Page to Stage." *Interdisciplinary Literary Studies,* vol. 7, no. 1, 2005, pp. 84–103.

McGilchrist, Iain. *The Master and His Emissary: The Divided Brain and the Making of the Western World.* Yale UP, 2019.

Menely, Tobias, and Jesse O. Taylor. "Introduction." In *Anthropocene Reading: Literary History in Geologic Times,* edited by Tobias Menely and Jesse O. Taylor. Pennsylvania State UP, 2017, pp. 1–24.

Morton, Timothy. *Ecology without Nature: Rethinking Environmental Aesthetics.* Harvard UP, 2007.

Munday, Anthony. "A Second and Third Blast of Retreat from Plays and Theatres (1580)." *Shakespeare's Theater A Source Book*, edited by Tanya Pollard, Blackwell Publishing, 2004, pp. 62–83.

Nellhaus, Tobin. "Introduction: Speech, writing, and performance." In *Theatre Histories*, by Bruce McConachie et al., general editor: Tobin Nellhaus, 3rd edn., Routledge, 2016, pp. 21–24.

Northbrooke, John. "A Treatise Against Dicing, Dancing, Plays, and Interludes, with Other Idle Pastimes (1577)." *Shakespeare's Theater A Source Book*, edited by Tanya Pollard, Blackwell Publishing, 2004, pp. 1–18.

Powers, Richard. *The Overstory: A Novel.* W. W. Norton & Company, 2018.

Puchner, Martin. "Performing the Open: Actors, Animals, Philosophers." *TDR The Drama Review*, vol. 51, no. 1, 2007, pp. 21–32.

Stevens, Anne H. *Literary Theory and Criticism.* Broadview Press, 2015.

Rigby, Kate. "Writing After Nature." In *Ecocriticism: The Essential Reader*, edited by Ken Hiltner. Routledge, 2015, pp. 357–367.

States, Bert O. *Great Reckonings in Little Rooms: On the Phenomenology of Theater.* University of California Press, 1985.

Thunberg, Greta. *No One is Too Small to Make a Difference.* Penguin Books, 2019.

van Dooren, Thom, and Deborah Bird Rose. "Storied-Places in a Multispecies City." *HUMaNIMLIA*, vol. 3, no. 2, Spring 2012, pp. 1–27. https://www.depauw.edu/humanimalia/issue%2006/pdfs/van%20dooren%20rose.pdf

Yusoff, Kathryn. *A Billion Black Anthropocenes or None.* University of Minnesota Press, 2018.

Part 5

Design and production

13 Eco-scenography and sustainable theatre production

David Fancy in Conversation with Tanja Beer and David Vivian

DAVID FANCY Let's begin discussing the challenges of considering the notion of eco-theatre, environmentally conscious theatre, climate crisis response theatre or climate positive theatre—whatever our favourite designation. Why has this not taken off sooner in areas of production, scenography and design? What are some overall perspectives on "sustainability" in these contexts?

DAVID VIVIAN How we teach production in North American is to run pretty close to the bone: the choices we make do not have a future planned into them. The adoption of sustainability models such as the seven-year model innovatively pursued in some Japanese communities (adopted from North American Indigenous culture) is certainly not often on the table for us in production pedagogy. Our decision-making process for pursuing sustainable frameworks is often absent. This is also true in our model of (usually unsustainable) professional practice. These things are related. For example, the question of sustainability comes up in the consumption and the reuse (or not) of materials, recycling through practice, just as does the question of sustainability of personnel in management and human resources contexts. In other words, questions of sustainability arise in the consideration of consumption of power, resources, water just as it comes up in the question of human resources and labour. And this has all been brought into focus with the pandemic at the moment. For example, within the Canadian context we have two professional design associations, one of these being APASQ (Association des Professionnels des Arts de la Scène du Québec), and another, an association of theatre designers working in English-speaking Canada, the ADC (Associated Designers of Canada). Because nobody has any work, there are no filing fees, no membership fees, there is no money to sustain the ADC for example, so we have recently voted to pursue what only a few years ago would be considered a contested path forward of radical change. This comes at a time when theatres are closed and possibly will be for a couple of years. This is having huge implications on employment of artists, the employment of production personnel; perhaps a little less on the administrative land-scape. Before the pandemic happened, we were having some serious conversations in the field of theatre design and the unsustainability of our practice from a professional perspective. Now it has become that much more real and necessary to negotiate something we have been quietly looking at from across the room.

The bigger picture here, and how this question of professional sus-tainability relates to important teachable questions of ecological sus-tainability more broadly in theatre training is this: these areas all relate to decisions we have made, decisions about consumption, betraying an absence of accountability regarding the consumption of resources that we have been working with throughout the 20th century. And now look at the situation we find ourselves in. Some areas of the world like California are burning up, Australia was last year, we have significant employment concerns in much of the first world. From the point of view of the models we have been working with I cannot imagine how artists and theatre-makers will persevere beyond the pandemic with-out there being significant, even radical, change. I remember a recent presentation by the head of the Canada Council for the Arts, who said, "You know you folks in 'theatre' are going to be the very last profession that re-opens to practise your craft sustainably"—sustainably in many senses of the word. This was a sobering observation.

DAVID F   This moment is one of what Noami Klein would describe as "disas-ter capitalism," in which overlapping crises, arguably issued forth from exploitive dynamics of capitalist production, create a moment of sig-nificant structural shift of even liberal democratic government's capac-ity to buffer their populations from capitalism's excesses, let alone protect the wider environment. And as you suggested there is a link between the ability to operate sustainably as a community with people who are involved with forms of creative labour and artistry. Before we started, Tanja, you described yourself as an "odd person out," with an interest in eco-theatre: can you talk about what you've perceived and experienced as being the barriers to the introduction of these types of thinking into what we might call more traditional theatre-oriented pedagogy, and the kinds of solutions you seek to find through your own interdisciplinary teaching and work?

TANJA BEER   I did my Ph.D. across architecture and theatre and now teach Interior/Spatial Design at the Queensland College of Art. In wider fields of design, ecological practice is seen as an opportunity rather than a constraint. It's all about innovation. Teaching sustainability in architecture, product and interior design is not questioned; if anything, it is embraced. I've never had a student say to me, "I don't want to do eco." It is absolutely not even a question for them: they are already liv-ing with such a strong awareness of the unsustainable reality that they find themselves in. They want to be part of the solution.

One of biggest barriers to sustainable theatre production is the fact that many people still see sustainability as a limitation—something that "limits high-quality aesthetics," is "tedious," "time-consuming," and "costly." It was not too long ago that I was told that, "Theatre and sustainability don't mix." Thankfully, the leadership of Greta Thunberg and others is finally hitting home.

DAVID F What kinds of curricular components and teaching strategies could you imagine being part of an education that might change this perception?

DAVID V I think that from a very pragmatic point of view I would be introducing students to projects like Julie's Bicycle in the UK and to the tool kits that they put together. This is a project and a series of measuring protocols that we're wanting to bring to our own practice here at the School of Fine and Performing Arts at Brock University where I work. Other similar sustainable theatre organizations and approaches are increasingly available. It is about teaching the students the necessity of a 360-degree consideration of their production concerns and aspirations, and bringing things right down to the nitty-gritty of all details— material, temporal, social—among the many. Let's take an account, let's model something, let's make some different choices, and then let's see how the results are produced as a consequence of our choices, and how we can achieve an agreed upon set of goals, maybe even discover new outcomes.

Another important consideration is that over the last couple of years the students we are bringing into the studio are very much attached to the idea that theatre can be a tool that will assist them to rebuild the world, to reorganize and reconceptualize the world. As such they situate themselves into different sorts of futures that we may not know yet, marked for example by reconciliation with our Indigenous neighbours, and of the learning that those of us who are not Indigenous can receive from them. The more that we can bring thinking about the role of the mother and the teachings of the grandmother from an Indigenous perspective into the teaching studio, the more we can connect such ideas to a reorganization of perception of what aesthetic value is and might be; to thinking about how we might participate differently in the creation of the aesthetic experience, and how we might differently position ourselves to receive and participate with it.

FROM THE PERSPECTIVE OF PROVIDING EXAMPLES, ONE OF THE MOST COMpelling experiences for me at the Prague Quadrennial in 2019 was at the presentation of the awards for performance design in space that featured the project from Hong Kong entitled *Theatre in the Wild*, a temporary site-responsive intervention in a rural landscape threatened by urban development. A young company of architects conceived of a small, transitional, transitory, portable recycled and recycling venue that could be easily transported. They had a limited time for their build, and the installation was conceived to be made out of cloth, pallets or other found recycled and reclaimed objects that they were bringing into spaces situated at the border of Hong Kong and Guangdong Province of the People's Republic of China. So, this was also a political gesture. This is taking the public out of the downtown core; a public that is obviously electrified by their need to respond to and maybe demonstrate against the changes with the transition of rule, taking it to more of a rural milieu. I wouldn't call it suburban, because these are sometimes more like Brownfield[1] spaces, for example, where they would set up, and that was a combination of aesthetic simplicity. I found it a very satisfying design concept; it had a really strong sense of the material they used and of its sustainability; it had social and political currency to it because of the relationship between China and Hong Kong. They were doing something that was meaning to be provocative and to be seen, and yet so discreet and with such a light footprint that there was great beauty to it.

So, to bring it back to the question of pedagogy: these are types of ways of making theatre that I would want to bring into our theatre-making experience with the students, and hope to maybe recalibrate institutional assumptions and practice by helping them understand why this could be vital and meaningful change going forward. This is particularly the case when some of our students come out of many positive high school experiences in lyric theatre and music theatre for example, and are used to investing a lot of energy in reproducing Broadway—a theatre production practice that is often very far removed from the considerations of environmentally conscious theatre.

DAVID F What would your curriculum look like, Tanja, perhaps implementing some of the things that you're doing in that crossover between architecture and theatre, and some installation pieces that you've done; how would they resonate with a kind of installation that David just described?

TANJA I teach "Sustainable Environments," which is a core subject for our second year Interior/Spatial designers, where students are introduced to both theoretical and practical perspectives on sustainability. We take students through a series of provocations each week that include questioning their assumptions of sustainability and sustainable design, and opening up new ways of thinking through the ecological worldview. We explore Indigenous knowledge systems, more-than-human perspectives, systems thinking, the circular economy, bio-inspired design, biophilia and regenerative development, and examine precedents across art, architecture, interior, installation and theatre design.

The course also has synergies with my forthcoming book, *Ecoscenography: An introduction to Ecological Design for Performance* (Palgrave Macmillan, 2021). The book applies broader notions of ecological thinking and practice to theatre production and reconsiders how we can realign ourselves with more biomimetic and circular ways of doing things.

A particular focus of eco-scenography is the making and unmaking of our practices, where all three stages of production—"co-creation" (pre-production), "celebration" (performance season) and "circulation" (post-production)—are equally valued.

DAVID F Interesting, as there's resonance here with some of the early performance studies in articulations of theatre ritual, or performance and ritual, with the warm-up phase, the presentational phase and the warm-down and cool-out phase. I'm wondering what things come to mind when we are looking at de-centring the *anthropos* in Western tradition—not that this is the only tradition worth consideration obviously—allowing the focus of theatre to shift from the supremacy of the human towards a consideration of complex interdependent and open-ended systems? I'm not talking about being nihilistic or misanthropic about human, but about creating space for the agency of other living things and nonliving things. How would we bring that kind of perspective into teaching in these areas?

DAVID V I can't offer a very illuminating answer to that, but just off the cuff I will offer that so much of our teaching and production work is currently based on existing texts, and that if we are mining from an existing and limited canon of plays, then the shift you're speaking of is not likely to happen. So, I think it is largely a question of mobilizing new, in-the-moment creative processes, and using those values as key objectives in order to structure new experiences. Various artists are doing this kind of work and are excellent teaching examples, such as Alexandra Lorde with Trigo Creative, whom I will be working with soon, and who presented recently to my first-year students. She literally exploded their minds with her perspectives on non-linear processes that seek to create points of escape beyond the Anthropocene.

TANJA I think you bring up an interesting point about how educators might ask students to respond to an environmental text to help bring ecological perspectives into their design thinking. Something I also find very interesting, working across architecture and theatre, is that architecture students are generally taught to respond to place. They are expected to consider the context and historical layers of the site, including the more-than-human communities that make up that place. Generally speaking, we don't teach stage designers to respond that way. We give them a theatre building or a play and say, "Design for that stage, for that piece." We don't necessarily expect stage designers to respond to the broader site of the theatre or to acknowledge the communities that make up that place. A key part of teaching sustainability in performance design is about opening our students up to wider perspectives of site and community.

DAVID V I agree, and when thinking of wider contexts I am observing, for example, that over the last couple of years a number of directors I have been working with have been very interested in patterns of migration and displacement of populations. This has significant impacts on contexts of sustainability. People are moving up from the African continent, across the Mediterranean and Europe. We are also looking at engineered migration: think of the New Silk Road, known as the "Belt and Road" initiative, the huge Chinese investment to re-establish trade pathways from the East to Europe. Here we are now living in the context of a pandemic where mobility is being reframed: those who are normally perceived as having privilege, for example, and have the ability to buy a ticket and get on a plane, are bumping up against significant restrictions, in a world that has also recently experienced significant displacement of populations, often immigrating to safe spaces away from social, political and economic environments put at risk in the face of the climate crisis, and even catastrophe. I'm interested in the lightness that this mobility invokes, and I'm thinking about designing through a certain aesthetic; perhaps qualities of a modernist point of view, where light and efficient approaches can help us sustain a lightness of footprints of time, space and cost in theatre production, for example. It's a really rich time to be evaluating and connecting so many forces contributing to the learning of our students.

DAVID F Let's finish then with some observations about students in particular. Some of the considerations that other contributors have put forward revolve in part around young people experiencing a particular form of affect that is a mix of climate grief and climate justice activism. What kind of holding strategies, teaching approaches and care are we going to need to help students deal with the significant psychosocial pressures they face, living as they will have to in the shadow of the climate crisis for their entire lives?

TANJA Extinction Rebellion immediately springs to mind, because of the way that we're seeing activism and theatre come together around the climate crisis. One of the reasons that Extinction Rebellion has been so popular is because it is so performative and scenographic. I would love to see more theatre and activism coming together in public spaces. I am passionate about finding opportunities to design exciting spaces for people from all walks of life; people who might not always have access to artistic experiences, and to make theatre in places where you might least expect it. And while we need to train students to work across traditional venues, I think an expanded, diverse and inclusive approach is integral for creating more sustainable employment opportunities for our graduates, now and into the future.

DAVID V I'm thinking in particular about the sense of loss and the sense of pain that our current student populations are experiencing, and their hunger—which they may not recognize, may not even know about—for spaces that provide them with a sense of healing or safety or spiritual wonderment. I'm not thinking about organized religion and their sites of worship per se, but of how space can nurture a certain amount of spiritual and healing experience for us. I'm thinking about the connection to a history of places where the practice of spirituality peels back veneers of material existence and consumption; of an examination of the capitalist systems that have brought us into the mess we're in, and students' hunger for something different, and that may remind some of us old enough to remember the hope and aspirations of post-Second World War 1960s culture. Our students are discovering this in a new way and making it their own. How can I assist them in designing space that situates and fosters hopeful futures, and that stimulates them in terms of their spiritual ecology? This may be born of a renewed theatre practice and it may be born out of the architectural design, or topographic or geographical and site-based scenographic practice. How can we encourage them to understand their connection to sun and moon cycles, for example, and why this matters to their creative and technical decision-making processes in consideration of the experience of the artists and the audience? How are we going to mentor them to celebrate their desire to make a difference in their chosen craft of theatre-making, and, given their drive to sustain different ways of being, to adopt and invent new approaches that propel them out of a long history of unsustainable theatre practice?

Practically speaking…

- Teach students to consider the relationships of sustainability between the use of resources, but also related issues of the sustainability of labour practices.
- Challenge the perception that sustainability represents a limitation, or indeed presumes one particular or restricted aesthetic.
- Connect sustainable practices to issues of Indigeneity and decolonization.
- Ensure that curricula include systems thinking, the circular economy, bio-inspired design, biophilia and regenerative development, and examine precedents across art, architecture, interior, installation and theatre design.
- Teach students about the ways in which designed space can nurture opportunities for spiritual and healing experiences.

Note

1 "A brownfield is a property, the expansion, redevelopment, or reuse of which may be complicated by the presence or potential presence of a hazardous substance, pollutant, or contaminant." United States Environmental Protection Agency. *Overview of EPA's Brownfields Program.* https://www.epa. gov/brownfields/overview-epas-brownfields-program#:~:text=Definition%20 of%20a%20Brownfield%20Site,substance%2C%20pollutant%2C%20or%20 contaminant (accessed 25 September 2020).

Epilogue

Theatre pedagogy and the climate crisis—a manifesto

Conrad Alexandrowicz, Mary Anderson, Gloria Asoloko, Lara Aysal, Tanja Beer, Soji Cole, Derek Davidson, Katrina Dunn, David Fancy, Dennis D. Gupa, Alexandra (Sasha) Kovacs, Rachel Rhoades, Kirsten Sadeghi-Yekta, Caridad Svich and David Vivian

We acknowledge and affirm the severity of the climate crisis, and the role that theatre education and training must play as part of the response to this human-made calamity. We believe that, since theatre can engage humans on corporeal, spiritual, intellectual, affective and social levels, the art form in all of its manifestations is uniquely positioned to engage the climate crisis. Despite these potentials, however, we also acknowledge that theatre, with its traditional focus on the human experience, can also be uniquely unwilling to acknowledge the anthropocentric origins and species-ist narcissism of the vast majority of its current expression. Similarly, while we affirm education's potential to help instigate radical futures marked by equity-oriented individual and collective enterprise, current global economic forces often reduce post-secondary education to the simple acquisition of skills required to survive in an alienating neo-liberal marketplace.

With these challenges very much in the forefront of our thinking, we advocate for the following agendas, priorities and commitments:

- Theatre educators must acknowledge, understand and respond via our pedagogy, to the range of specific stressors to which our students are subject, including but not limited to: feelings of grief and loss about the climate emergency; social alienation resulting from economic systems predicated on pervasive productivity, vicious competition and lack of predictable financial security; and intersecting minoritizations related to lived experience of racialization, gender-based violence, able-ism and other similar forms of violence. As educators we need to understand—and help our students to understand—that these forces are intimately related to the ecocidal and matricidal tendencies motivating the long-term plunder and desecration of the planet's other-than-human entities.

- With a view to generating a strong sense of agency, solidarity, accountability and purpose in our students, we advocate for youth-led cultural production that responds as directly possible, without being entirely reduceable, to student articulation of priorities in their study, leadership and creation. We do not perceive such a commitment to challenge our frequently hard-won experience and expertise, but instead as a means of best sharing our experience in an accelerated and equitable fashion, as befits the urgency of the climate crisis.

- We affirm the value of using the teaching and learning of theatre, and the study of its potentials, no matter from what vantage in the art form, for the purposes of community building, community engagement and the pursuit of intergenerational exchanges, as means of building the collective solidarities required to mount a substantive and effective climate justice movement in response to the climate crisis.

- Recognizing and honouring the many differences between our sub-disciplines within theatre, as well as the many different ways in which theatre stages a complex set of affective, cognitive, aesthetic, material, social and other interventions in the milieu in which it is pursued, we embrace a fulsome ecology of strategies and tactics in the pursuit of climate justice for all of earth's creatures and environments.

- We do, however, acknowledge that the more ambitious and radical our goals and agendas, the more previously radical objectives may appear moderate, and by extension more palatable to the political centre, and therefore the more we can aspire to accomplish.

- We affirm and stand in solidarity with global anti-colonial efforts and movements that seek the indigenization of teaching and learning spaces. While not in any way discounting the many affirmative and equity-oriented modes of thinking that have emerged from Western conceptualization and inquiry, we nonetheless advocate for the significant role that hitherto suppressed Indigenous ontologies and epistemologies can play in redressing the conceptual imbalances fuelling the climate catastrophe.

- Indeed, we recognize that theatre cultures, histories and the stories that theatre continues to foreground are frequently marked by the kinds of misogyny, racialization and other kinds of minoritizations that intersect with the ecocidal imperatives generating anthropogenic climate change.

- We are deeply sceptical of "green growth" or "sustainable development" agendas that assume an increase of any national gross domestic product to be part of the solution to the climate crisis. We commit to integrating our reimagined and reconfigured theatre pedagogy with the most effective critiques of unbridled and metastasizing capitalism.

- We commit to advancing a rich and varied eco-literacy within our theatre programmes, drawing on notions of an affirmative planetarity, biophilia and multi-modal understanding of sustainability. We challenge ourselves and our colleagues to understand that the distinction between "nature"' and "culture" is a false one, and to pursue our practical, creative and conceptual work in classrooms, workshops and studios with this in mind. We endorse the understanding of a deeply imbricated and entangled relationship between the human and the other-than-human, and work to problematize reductive perspectives on representation and mimesis that are integral to many theatre cultures globally.

- While not committing to one single conceptual mode of challenging anthropocentrism, we seek to advance a range of (often competing) critical posthumanist and philosophical perspectives in our theatre and performance studies. Far from advocating an anti-human nihilism, we continue to celebrate the value and complexity of infinite types of human experience, but not at the expense of all other creatures and entities on the planet, or of the planet itself. Indeed, we celebrate the role of thought, reflection, consideration and artistic expression, while seeking to understand how aspects of these endeavours—often considered to only occur within human cultures—may in fact be distributed widely across a range of phenomena. In short, regardless of our conceptual agendas, we seek not to suppress human *singularity*, but instead to celebrate it without lapsing into the illusion of human *supremacy*.

- Since the human failures that created the climate crisis result in part from a crisis of the imagination, we commit to helping our students pursue the energetic release and harnessing of the theatre's most precious constitutive capacity—that of creativity—to generate the wildest, subtlest, most far-reaching visions, dreams and manifestations of what a post-anthropocentric planet might look like. We recognize that these visions may just as likely happen in the acting studio, the applied theatre community placement project, the carpentry shop or wardrobe and so forth, as anywhere else. We celebrate the intra-disciplinary resonances to be generated by affirmative mutual solidarity across our many sub-disciplines.

- Despite the urgency of the climate crisis, we vigorously defend approaches to theatre pedagogy that privilege curiosity, discovery, open-ended inquiry and a commitment to the acceptance of failure as an integral part of the process of learning, and of theatre-making itself. We advocate love, loving-kindness, as well as self- and collective acceptance, as being integral aspects of the kinds of dependencies and interconnections we seek to engender between ourselves, our students, our human communities and the planet's many other-than-human creatures and entities. We advocate acknowledging and supporting our students' healing, wellness, and their journey out of trauma of all kinds, as a means of facilitating their planetary solidarities.

- We do not expect ourselves, or our students, to remain unchanged by the agendas articulated here: we embrace individual, collective and planetary transformation.

Note: If readers are coming to this statement outside of the context of the entire volume, entitled *Theatre Pedagogy in the Era of Climate Crisis*, please note that each of the individual chapters in the volume concludes with a "how-to" section that contains broad ranges of often immediately actionable steps to be taken across many areas of theatre education and training.

Index

Note: Page numbers in *italics* refer to content in figures.

Printed in the United States
by Baker & Taylor Publisher Services